CAVE INVESTIGATION

Edited by **Savaş Karabulut**
and **Mualla Cengiz Cinku**

Cave Investigation
http://dx.doi.org/10.5772/65541
Edited by Savaş Karabulut and Mualla Cengiz Cinku

Contributors

Jan Zukal, Hana Berková, Jiří Pikula, Hana Banďouchová, Veronika Kováčová, Gáspár Albert, Aurelija Rudzianskaite, Serbülent Paksuz, Manuel Matias, Fernando Almeida, Nuno Barraca, Biswajeet Pradhan, Mohammed Oludare Idrees

Published by InTech
Janeza Trdine 9, 51000 Rijeka, Croatia

Publishing Process Manager Romina Rovan
Technical Editor SPi Global
Cover InTech Design team

First published July, 2017
Printed in Croatia
Legal deposit, Croatia: National and University Library in Zagreb

Additional hard copies can be obtained from orders@intechopen.com

Cave Investigation, Edited by Savaş Karabulut and Mualla Cengiz Cinku
p. cm.
Print ISBN 978-953-51-3331-5
Online ISBN 978-953-51-3332-2

For EU product safety concerns:
IN TECH d.o.o., Prolaz Marije Krucifikse Kozulić 3, 51000 Rijeka, Croatia,
info@intechopen.com or visit our website at intechopen.com.

Meet the editors

Savaş Karabulut graduated from the Istanbul University, Department of Geophysical Engineering, Turkey in 1997. He received his master's degree in 2001 and PhD degree in 2012 from the Istanbul University, Department of Geophysical Engineering. He worked as a research assistant and assistant professor doctor at the Istanbul University Department of Geophysical Engineering for 15 years. His studies have focused on engineering seismology and soil engineering problems in Turkey. His current research is focused on geophysical applications applied on caves in Turkey, especially on the Yarımburgaz Cave (Küçükçekmece, Istanbul). He is the author and co-author of several books, articles, and conference papers. He is a member of UCTEA, The Champer of Geophysical Engineering, European Geoscience Union, American Geophysical Union, and Earthquake Engineering Association of Turkey.

Assoc.Prof. Dr. Mualla Cengiz Cinku earned her Ph.D. from the Istanbul University, Department of Geophysical Engineering in 2004. She is currently working as an associate professor doctor at the Istanbul University Faculty of Engineering Department of Geophysical Engineering. Her major research interest is paleomagnetism and rock magnetism. Her current research is on the cave detection with geophysical techniques. She has published many articles and conference papers. She is a member of UCTEA, The Champer of Geophysical Engineering, European Geoscience Union, and American Geophysical Union.

Contents

Preface

What is a cave? [1] Walker (2008) defined caves as: "Imagine a place that's dark and cool. That icy air chills your body on a hot summer day. A river flows out of rock. Stone icicles drip water. A group of bats suddenly appears. All of these things can happen in a cave."

Caves are natural underground holes finding openings on the surface. Almost every country in the world has caves. Most caves are small and form in karsts taking long time. They have many chambers, entrances, and interconnecting passages. The largest single cave passage was discovered in Vietnam, the Son Doong Cave, with a length of 80 meters and a depth of 4.5 kilometer.

"Speleology" is the scientific discipline used for cave investigation. The first cave studies were undertaken by French archeologists, which later on spread worldwide. Cave studies that find scientific exploration by using different disciplines are first applied on determining habitats of early humanity and afterward focused on their archeological and biological properties. Throughout time, a lot of effort was spent on cave art. Besides, the use of subterranean rivers in caverns and their importance as roost sites for bats and species are focused as primary research studies. Caves have become tourist destinations because of their natural and cultural heritage, while the unique fauna and flora in caves are required to be managed and protected. Its chemical, physical, and biological effects are important for the living entities, for instance, inhabitants of colonies of bats or species.

Today, evidence of past cultures can be found in caves by means of artworks and archeological remains around the world. The size, elongation, and the width/length of caves began to be determined by archeologists and speleagloists through excavation and the use of lasermeter. Each year, several teams of scientists undertake major expeditions to discover new caves, while the formation of caves serves scientific problems that need to be clarified.

Nowadays, the past environment can also be explored by series of geophysical measurements in order to study the stability of galleries, foundations, structures, etc. Most common methods of direct measurements are known as probing, sounding, excavation, test pits, and trenches. These methods are destructive and costly and provide information depending on the subsurface conditions over large areas. On the other hand, geophysical techniques are nondestructive and may be applied to mapping sinkholes, fractures, and continuations of caves with high accuracy and efficiency.

This volume provides references of the results of methods and extensive case studies applied on important caves in Europe and Asia carried out by several researchers. The results will help many researchers in geoscience, archeology, climatology, and biology. In this volume, three sections discuss the following topics: (1) Geomorphometric Analysis of 3D Cave Mapping, (2) Bats in Caves, and (3) Geophysical Application for Cave Detection and Geoecological Assessments of Water Chemistry in Karst Aquifers. A brief information of each research given in this volume is provided below.

The geometric parameters of cave morphology using terrestrial laser scanning (TLS) were applied on Gomantong Cave in Malaysia Borne in the chapter "Characterization of Macro and

Micro-Geomorphology of Cave Channel from High Resolution 3D Laser Scanning Survey: Case Study of Gomantong Cave in Sabah, Malaysia" by Idrees and Pradhan. The identification of macromorphological and micromorphological features of the Gomantong Cave was achieved by 3D models.

In the study "Aspects of Cave Data Use in a GIS" by G. Albert, the geographic information system (GIS) was used to analyze cave survey data on how to work with archive and new survey data and how to handle maps, scans, and sampling data. The chapter focused on the importance of procedures of data management, quality control, and automation of cave survey data.

Caves as hibernacula and the effect of temperature and bat hibernation are discussed in the study "Bats and Caves : Activity and Ecology of Bats Wintering in Caves" by J. Zukal, H. Berková, H. Banďouchová, V. Kováčová, and J Pikula. The authors showed the importance of the flight activity at the cave entrance and the bat movement activity inside the cave.

In the study "Important Caves in Turkish Thrace for Bats: Dupnisa Cave System and Koyunbaba Cave," by S. Paksuz, two important caves in Turkish Thrace showing different microclimate characteristics were discussed as a result of the species composition, colony structure, seasonal population dynamics, roosting habits, and roosting requirements of bats. It indicated that the protection of Dupnisa Cave System and Koyunbaba Cave is very important for the future of bat populations in the region.

A review was present of the results from various investigations on drainage, rivers, sinkholes, and ground water in the agricultural area of North Lithuania karst region in the study "Investigation of Water Quality in the Agricultural Area of Lithuanian Karst Region" by A. Rudzianskaitė. It showed that chemical composition of research water depended on the type of soil, meteorological conditions, and land use.

In the study "ERT and the Location of Mining Cavities in Anisotropic Media: A Field Example," by M. Matias and F. Almeida, a field study was applied on an old mining area to demonstrate that 2D resistivity data—Electrical Resistivity Tomography (ERT)—can be strongly affected by local anisotropy that masks the presence of cavities in ERT data modeling. The authors addressed the importance of the field survey design.

Reference

[1] Walker SM. Caves. Minneapolis, USA: Lerner Publications Company; 2008. ISBN-13:978-0-8225-6734-9

Dr. Savaş Karabulut
The Chamber of Geophysical Engineering Branch of Istanbul
İstanbul, Turkey

Assoc. Prof. Dr. Mualla Cengiz Cinku
Istanbul University Faculty of Engineering
Department of Geophysical Engineering
İstanbul, Turkey

Section 1

Geomorphometric Analysis of 3D Cave Mapping

Chapter 1

Characterization of Macro- and Micro-Geomorphology of Cave Channel from High-Resolution 3D Laser Scanning Survey: Case Study of Gomantong Cave in Sabah, Malaysia

Mohammed Oludare Idrees and Biswajeet Pradhan

Additional information is available at the end of the chapter

http://dx.doi.org/10.5772/intechopen.69084

Abstract

Three-dimensional documentation of hypogene cave morphology is one of the major applications of laser scanning survey. This chapter presents applications of terrestrial laser scanning (TLS) survey for analyzing endogenic cave passage geomorphologic structure and morphometry using 3D meshing, high-resolution 3D texture modeling for geovisualization, and its potential for cave art documentation. To achieve this, multi-scale resolution 3D models were generated; one using the mesh model for macro-morphological analysis and the other with the full-resolution scan to produce high quality 3D texture model for identification of micro-morphological features. The mesh model of the cave makes it possible to analyze the general shape, distinguish phreatic tube from post-speleogenetic modified conduits and carry out morphometric measurements including the cave volume and channel surface area. The 3D texture model provides true to live visualization of the cave with exceptionally high level of accuracy and details that would be impossible to obtain with direct observation by visiting the site or from the mesh model. The model allows discerning different speleogenetic phases, karstification processes and micro-morphologies such as wall and ceiling seepage, hanging rocks, fractures, scallops, ceiling flush dome, pockets, bell-hole and avens. Also, the texture model permits identifying cave arts and engravings along the passages

Keywords: virtual geomorphology, texture model, cave, passage failure, terrestrial laser scanning, Gomantong cave

1. Introduction

Caves are natural resources of complex dimensions that require intelligent planning and management for sustainability. They are naturally formed caverns that have played diverse roles for the human and animal communities. Investigations have revealed primordial use of caves on a short- or long-term basis depending on the nature of activities they are used for or the circumstances that necessitate the usage [1]. Caves have been used by man as shelter and for protection during wars. In many cultures, people bury their dead ones in caves occasionally, with their belongings and gift to help them in their spiritual journey. Man has used and continued to use caves for ritual and religious functions till this present time. Studies have also confirmed that caves were used for industrial purposes. For example, relict ceramic and metal deposits found in the French Bronze Age cave, Les Fraux, indicate that the ancient bronze workers used the cave as "industrial" workshop [2]. It is also on record that Bedeilhac cave, in the French Pyreness, was used as aircraft factory during the post-Industrial Revolution [1].

Human beings are rarely the first to explore caves. Caves are favourite natural habitat for varieties of animals, such as pigs, primates, elephants, hornbills, bats, birds, reptiles, amphibians and other organisms, some of which are endangered. Evidence of visitation of animals to some caves has been found in their dung and bones. For example, skeleton of a young elephant was found in a cave in Pahang, Malaysia [3]. Likewise in Kitum caves, Kenya [3] reported the regular visits of elephants to eat minerals from the rocks. For over 600 years, the Great Niah and Gomantong caves in Sarawak and Sabah, Malaysia Borneo, have been famous for sheltering large colonies of bats and swiftlet birds [4].

This age-long interaction of man and animals in caves has resulted in enormous treasures within the confinement of the environment, albeit, it has also endangered the delicate cave ecosystem. In the interest of exploration, exploitation and preservation of the cave environment, man has devised several methods of surveying cave and tracking the resources therein in line with advances in technology of surveying instrumentation. Historically, earliest cave explorers used sketches to communicate or report their observations and findings in the cave. As time passed by, traditional instruments such as compass, clinometers and theodolites progressively became handy for cave mapping. By the turn of 1961, light amplification by stimulated emission of radiation (laser) began appearing on commercial markets for medical application and ranging. Subsequently, rangefinder and total station emerged with laser-ranging capability, collecting single-point xy(z) data at a time. Then, cave mapping became much easier and accurate than with compass and inclinometer; however, cave maps were still delivered in 2D space.

Today, sophisticated light-scanning (or laser-scanning) sensors that are capable of collecting several thousands to million points per seconds have taken over from the traditional methods, delivering caves in their true 3D geometry at unprecedented level of accuracy [5]. Terrestrial laser scanning (TLS) survey has proven positive results in different applications such as archaeological investigations, 3D modelling and visualization, geomorphological analysis, passage stability and risk assessment, biological inventory and so on. A comprehensive review of the journey of laser scanning to caves including method, progress in hardware and software development for data collection and processing can be found in Ref. [6].

Awareness of the wealth of information conveyed in laser-scanning data and the significant role high-resolution 3D documentation can play in deepening the complete understanding of the cave motivates this present investigation. We propose that 3D texture model can offer an excellent platform for geomorphological investigations with high level of accuracy through virtual cave tour. So, building on the foundation of Ref. [7], this study experiments virtual geomorphological analysis with laser-scanning data collected in the less-accessible Simud Putih. The aim is to generate multi-scale resolution 3D models. One, using the 3D mesh model for macro-morphological analysis and the other, with the full-resolution scan to produce texture model for identification of micro-morphological features. The investigation revolves around three applications with the following objectives:

- Modelling the cave in 3D space at different resolution scales for analysing endogenic passage geomorphologic structure and morphometry.
- Identification of micro-geomorphological features of the floor, wall and roof through virtual cave exploration with high-resolution 3D texture model and its potential for cave art documentation.

On this note, this chapter on "Characterization of macro and micro-geomorphology of cave channel from high-resolution 3D laser scanning survey: a case study of Gomantong cave, Sabah (Malaysia)" presents the approach employed for the multi-scale resolution 3D modelling of Simud Putih and the different applications they were used for. The structure of the chapter is organized thus: following this introduction, the second part describes the location of the cave with emphasis on the geographic location, geology and speleogenesis. In the third part, the general methodology with terrestrial laser scanning, including (i) planning, (ii) 3D laser scanning survey and (iii) point cloud processing, is sequentially articulated. Other specific data-processing tasks for the target applications will be done within the context of the individual subsection. The fourth part outlines and discusses the results of the case study and their applications. The last part is the concluding remarks and the future research outlook.

2. Geological setting and study area

Geographically, Gomantong cave is located in a tower-like limestone outcrop that projects approximately 300 m above a floodplain on longitude 118° 04′ E and latitude 5°32′ N within the Sabah forest reserve [7]. Travelling by road, the hill lies some 31.4 km south of Sandakan and about the same distance to the east of Kota Kinabatangan, the state capital of Sabah, Malaysia (**Figure 1**).

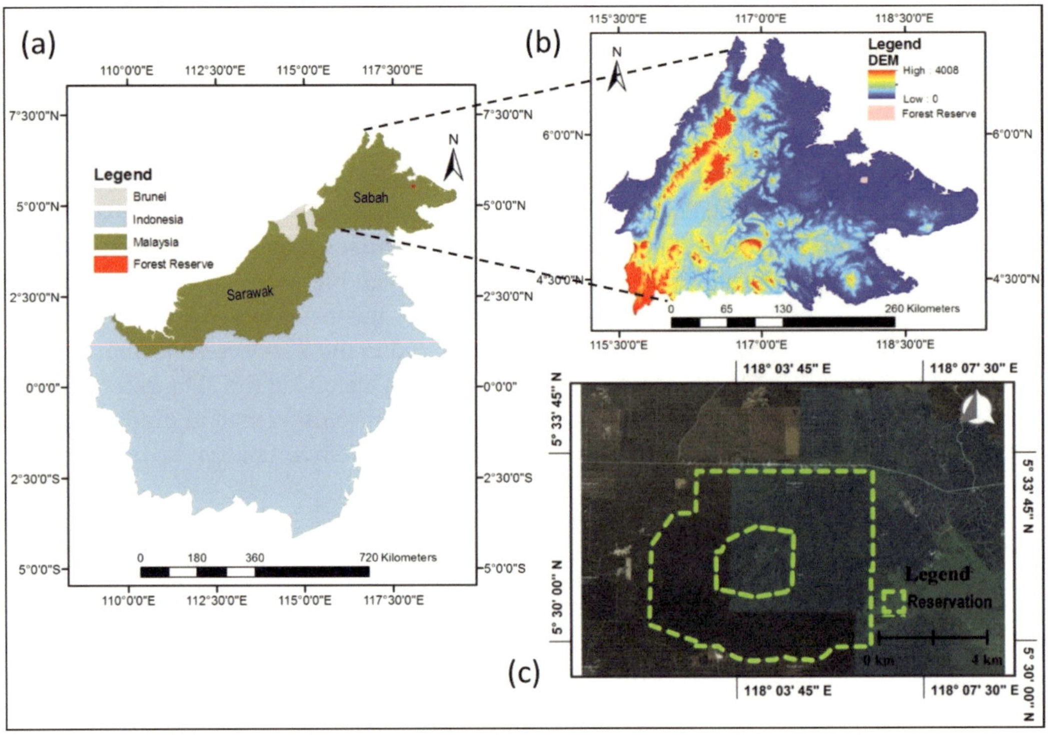

Figure 1. Location of Gomantong cave in (a) Borneo Island, (b) ASTER-GDEM of the state of Sabah (Malaysia Borneo) and (c) Google Earth image of Gomantong forest reserve with the hill at the centre.

From geological point of view, the limestone contains dense lithic fragments of grey sedimentary rock rich in organic matters deposited in laterally thin and well-defined layers that lie uncomfortably on the underlying Labang bedrock [8]. The layers alternate with bed of thin grey-green fossil and sometimes intercalated by sandstones and shales of the Oligocene Labang formation. Evidence from nanofossil dating of samples taken from marls, mudstones and on benthic foraminifera across the region estimates the formation age at about 23 million years ago—between late Oligocene and early Miocene [8, 9]. The detritus particles are believed to have recrystallized during the Oligocene sediment formation, resulting in varying sizes of the crystal structure in the uplifted rock.

In terms of speleogenesis, the cave is believed to have started immediately after the uplift, compression and folding of the rock during the early Pliocene which was followed by dissolution of soluble rocks along bedding planes and joints [7, 9]. Gomantong cave system consists of two major passages, one above the other [10]. The lower cave, locally referred to as Simud Hitam, provides entrance along one of the major faults that opens to the base of the hill and coincides with the ground level of the bank of a small stream nearby. But, for the upper level cave (Simud Putih), the access is situated some 85 m above the floor of the opening to the lower cave [10].

Like every other cave, this one is dark and humid, providing an excellent microclimate habitat that is suitable for swiftlet birds and bats in their millions to share [11, 12]. In addition, the cave provides naturally confined speleogenetic setting that makes it particularly attractive. Swiftlet birds and bats contribute significantly to ecosystem engineering, tourism and economy of the state of Sabah [13]. However, there are indications of threat of potential collapse of the cave due to natural and anthropogenic causes [7].

The danger posed by over-exploitation to the survival of the Gomantong cave system requires holistic approach to confront the multifaceted challenges resulting from the activities of man, animals and nature combined. For example, the cave passage of Simud Hitam has been extremely affected by biogenic corrosion. Equally, the cave wall and roof are frequently hacked to provide support for hanging ladders through which edible nests are reached. All these, in addition to natural geologic processes, weaken the structure and thereby accelerate the rate of deformation in the cave. Since the last six centuries, the cave has been one of the focal references to edible bird-nest (EBN) harvesting in southeast Asia [7]. While the business flourishes, enriching the economy of the state and her indigenous residents, the survival of the cave itself for life-long production of this lucrative commodity has been largely neglected, though unintentionally [4], despite the vicious impact of biologically induced alteration of the cave passage. In this regard, provision of adequate information about the internal structure of the cave will be of paramount assistance to decision makers and cave managers on the best way to protect the cave and its resources.

Recent investigations by Refs. [7, 10] show that little is known about the cave structure, and particularly the geomorphology, thereby throwing an open challenge to the research community on the urgent need for deeper exploration. Nevertheless, stakeholders were sufficiently alerted on the danger of continuous deformation of the cave cavity should biogenic decay persist [7]. However, the message could not be clearly conveyed with respect to the cave geometry to the stakeholders who, mostly, are non-technical professionals. The reason being that the data were collected using traditional survey method, and the information presented in 2D, making it impossible to visualize the enormity of the biogenic activities in the cave.

Another concern is that the foregoing investigation was carried out in Simuh Hitam (lower cave), near the entrance where daylight penetrates to illuminate the cave. Thus, the internal geomorphology including the ongoing deformations of the cave cannot be described. Moreover, the distance of the cave roof from the floor is far and barely visible to an observer from the latter. The second study [10] utilized 3D scanning; nonetheless, only the basic quantitative measurements (length, area and volume) were extracted. The lower cave has always been given more priority in terms of research mainly because of its ease of access. This accounts for why most of the studies about the cave system have focussed on the lower section while the upper cave chamber has not been widely explored. For that, this present study centres entirely on the upper cave.

3. Materials and method

The overall methodological workflow used in this study is presented in **Figure 2**.

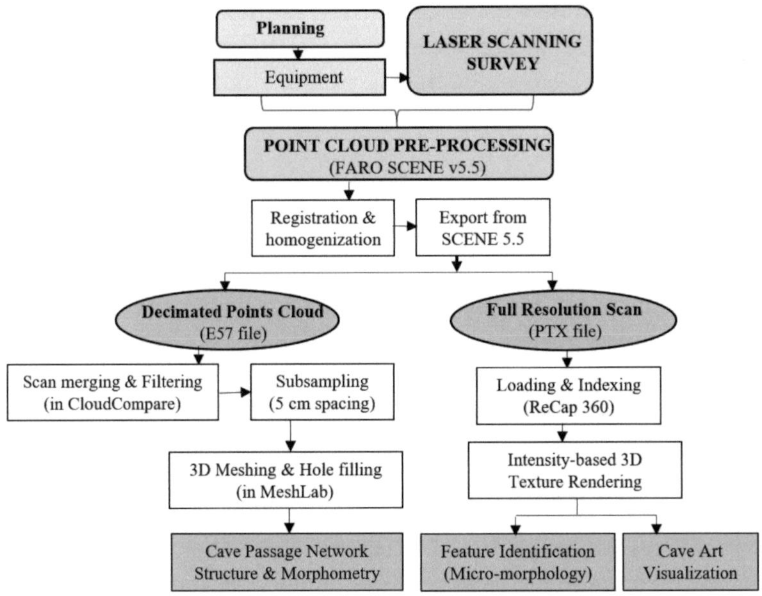

Figure 2. Data processing and analysis workflow.

3.1. Terrestrial laser scanning survey

The procedure presented here follows the established workflow for point data collection and processing: planning, field work and point cloud processing [6]. In July 2014, FARO Focus3D was used to scan Simud Putih to complement the 2012 scanning expedition in Simud Hitam [14]. The scanner employs the phase-shift scanning technology with exceptional distance accuracy up to a scanning range of 120 m. Like every other scanner, Focus3D measures distance to a target by comparing the difference in pulses of emitted and reflected wavelengths. The use of this particular scanner was evaluated on the basis of its technical suitability (**Table 1**). In the market today, Focus3D is the lightest and smallest high-speed scanner designed for outdoor and indoor applications [6]. The size and weight (~5 kg) of Focus3D, high accuracy, rapid rate of data collection and extended battery life of 4.5 h make it a suitable choice for high-precision survey in caves.

Planning is necessary to increase the efficiency of data collection and processing manoeuvrings for high-quality point cloud and 3D reconstruction [15]. Therefore, ahead of the field work, the cave was visited during which scanning positions were strategically predetermined to achieve good overlapping areas. Sufficient overlap between adjacent scans increases the quality of registration process and also provides complete data coverage. Moreover, distance between scanning positions and between the cave wall and the instrument was kept roughly at regular intervals to ensure consistent scan resolution.

During the data collection, a total number of eight spherical targets were used to support scan registration process. The targets were placed around the scanner field of view such that at least three or more can be visible in two adjacent scans for high-quality registration. Where

Feature	Specification
Manufacturer	FARO
Model	FARO Focus^{3D}
Range	0.6–120 m indoor or outdoor with low ambient light and normal incidence to a 90% reflective surface
Ranging error[2]	±2 mm
Measurement speed	122,000/244,000/488,000/976,000 points/s
Field of view	Vertical/horizontal: 305/360°
Weight	5.0 kg
Size	240 × 200 × 100 mm^3

Table 1. Technical specifications of FARO Focus3D terrestrial laser scanner.

reference to national or global coordinate system is required, the target positions are measured accurately using theodolite, total station or a combination of GPS and total station [16, 17]. Advances in hardware and software development have provided alternative scan registration procedures such as cloud-to-cloud and automated registration in target-less mode. Nevertheless, the need for target-based scan matching cannot be overlooked particularly in cave interior that is characterized by irregular formation, difficult access and unfavourable working environment. On each station, scanning was done at one-fourth scan resolution mode which provides 244,000 points/seconds and x-y point cloud spacing of 12.5 ± 2-mm ranging error at a distance of 20 m [14]. Overall, a total number of 203 scans were collected during the two weeks' field work that resulted in over 6.09 billion points of ~38.0 GB. Out of 203 scans, only 99 scans that represent the main passages of the upper-level cave were further processed and used for this study.

3.2. Point cloud processing

Since the scans were collected from different positions, it was necessary to align the individual scan into one set of point cloud. This was done using the proprietary SCENE-5.5 software (www2.faro.com/downloads/training/Software). For high-quality registration, semi-automatic target-based point correspondence utilizing the reference sphere placed within scan areas was used. The method employs the iterative closest point (ICP) [18] algorithm which automatically detects at least three artificial target points that are visible in successive adjacent scans to compute accurate transformation parameters. The process was further improved by activating auxiliary sensors (inclinometer, compass and altimeter) to speed up the correlation of the individual scans. The emerging trend in the new generation of scanners such as Focus3D x330, x130 and Z+F Imager 5010X is the evolution of multi-sensor scanning hardware including GPS receivers, inclinometers, inertial measurement unit (IMU), compass, powerful imaging cameras and height compensator that enhances the quality and accuracy of data collection and processing.

This particular study does not require reference to national or global coordinate system; hence, the scanner's local Cartesian system was sufficient. Usually, the first scan station is taken as the origin where the instrument's plumb line provides the z-axis [15], and the horizontal line is defined by the scanner's internal compass which facilitates aligning scans to geographic direction. The registration quality produced root mean square error of 0.0044 RMSE, equivalent to 4.4 mm. The registered point clouds were exported from SCENE software as an individual file in ASTM E57 file format for 3D imaging data exchange [19] and PTX file format, respectively. The E57 file format offers efficient storage and data compression that facilitates ease of 3D data transfer across different platforms better than ASCII-based point cloud exchange format like PTX. PTX file preserves all of the information from the original scan (scanner location) plus additional registration information (transformation matrices). The drawback of PTX is that it requires large memory space to keep the information. The aligned scan yielded point spacing of < 6 mm in both the horizontal and vertical axis.

For 3D reconstruction and rock structural analysis, processing the full-resolution scan is not computationally efficient [5]. Therefore, the E57 format point cloud dedicated for these applications was decimated by sampling every seventh point to optimize processing time based on the suggestion of Ref. [5]. In CloudCompare [20], a free and open source software, the decimated scans were merged to provide a single discontinuous point cloud of the cave. Thereafter, the matched point cloud was filtered to eliminate noise, while other points belonging to features outside the cave were manually deleted. Finally, the cleaned point cloud was spatially sub-sampled using horizontal and vertical distance spacing of 5 cm between points for efficient mesh generation. Further, application-based data processing and analysis using either the decimated or full-resolution scan point cloud are discussed in section 3.3.

3.3. Application-specific data processing

3.3.1. 3D Meshing and passage geomorphology

The sub-sampled point cloud in Section 3.2 was used for the 3D surface reconstruction. Point cloud comprises a set of dense 3D xyz points of the surface, but, in reality, the discrete point data do not offer sufficient description of the cave structure. 3D mesh generation was done in MeshLab 1.3.4BETA [21] to reconstruct the cave in its true 3D geometry. MeshLab is an open source package for point cloud processing, advance mesh generation and editing. A number of surface reconstruction approaches are available in the software but the widely used Poisson surface reconstruction approach was applied in this present chapter. The algorithm utilizes the sampled points from the surface and the gradient of the indicator function (a normal field) to build surface. The gradient of indicator function was derived by computing normals for the point sets using 10-point neighbourhood kernel. The procedure converts the point normal into a solid 3D surface by defining a scalar function that best matches the vector field to build a fitting isosurface [22, 23]. The 3D meshed surface was achieved with octree depth of 10 comprising 825,451 vertices and 1,650,604 faces. Octree depth parameter sets the depth of the octree used for extracting the final surface. Increasing the octree depth parameter increases the precision and level of details (faces and vertices) of the surface; however, compromise had to be made among level of details, visual reality and real-time interaction with the model. The final mesh model was used for analysing the general cave passage structural morphology and morphometry and also for the sectional drawing.

3.3.2. *Textured 3D modelling for detail micro-morphologic and cave art identification*

Documenting micro-morphological features requires high-resolution textured 3D model. Advances in computational efficiency of point cloud processing software like Autodesk ReCap and its textural 3D rendering capability will rapidly revolutionize cave investigation. Despite the tremendous success achieved with TLS survey in many notable caves around the world, photo-realistic 3D rendering of cave has remained unattainable. Besides the limitation of the existing point cloud processing software in handling large volume of data, overcoming the problem of total darkness has been a subject of intensive research [24–27]. Today, archaeological documentation using 3D scanning is considered the best method; nevertheless, the use of two independent sensors, laser scanning and digital photos throws up additional challenge. The difficulties include managing illumination variation, finding exact match between the images and geometric model and handling the volume of data acquired [27]. Besides, processing the data involves several manual operations.

As an alternative approach to RGB (red, blue and green) photos, the laser intensity value attached to each point was used as a substitute for the texture-based virtual reality rendering. In this study, we used Autodesk ReCap360 student licence package (http://www.autodesk.com) [28]. ReCap360 is one of the families of Autodesk products specifically designed for handling massive cloud point and photo dataset. It is innovative 3D modelling and reality capture software that create complex 3D model with real-world visualization by combining laser scans and photos. The package supports a number 3D file formats including CL3 (Topcon), E57, FLS (Faro), LAS, PCG, PTG (Leica), PTS, PTX, RCS, TXT, XYZ and ZFS (Zoller+Fröhlich). Moreover, it computationally optimizes input files by indexing to its native RCP file format for efficient processing, modelling and interactive visualization. The model allows complete natural feel of the rock such that all details of the irregular surfaces including rock engraving and artworks in their true 3D geometry are captured.

Understanding cave use, especially in the past, requires detailed true-to-life perception of the space in all dimensions including the morphogenesis of the floor, ceiling and wall caused by geological and, at times, biological factors. For Gomantong cave, no reference is found in the literature indicating cultural utilization of the subterranean space in the past outside nest harvesting. This aspect of our study aims at identifying micro-morphologic features and relict antiquity from the high-resolution 3D texture model that could provide insight into factors that influenced passage modification and enhances our knowledge of probable age-long use of the cave.

4. Discussion

4.1. Cave passage network structure

One of the main objectives of this study is to provide interactive digital representation that permits visualizing the entire cave network morphology in 3D space, and also, to allow extraction of features in their geometrically correct dimension without visiting the cave. The dipping limestone produced a particular style of cave that can only be comprehended in digital model. Meshing produced cave surface that provides exact representation of the arrangement, structure and irregularities (**Figure 3**). The cavity comprises connected passages geographically oriented to different directions. For specific description of cave channels and their morphological characteristics, the passages are classified into six segments (entry, chamber, hall and shafts 1, 2 and 3) according to their global orientation.

Meshing provides the structural construct of the pattern of the cave network. Externally, the plan view of the cave (**Figure 3b**) presents a system of network connectivity that bears a resemblance to typical Uzi Galil sniper rifle with each component (handle, butt, trunk, magazine and precision telescope) forming different channels, for example, entry passage, chamber, main hall, shafts 1 and 2 and shaft 3 in that order. The primary passage to the main hall starts off vertically downward to about 29 m at a very steep slope and then trends east-west, parallel to the main chamber, to join the hall. The chamber also connects the main hall at the upper-most end in the northern side through an orifice that dips west-east. This main hall is a large cavern that runs NE-SW, connecting the SE-NW trending shafts (1 and 2) and the inclined N-S shaft 3 (**Figure 3**).

The model allows cave interior visualization at the coarse level, while also permitting the extraction of dimensional data with detailed geometric definition and precise graphic description of the cavities (**Figure 4**). 3D mesh gives a feel of the general complex morphogenetic

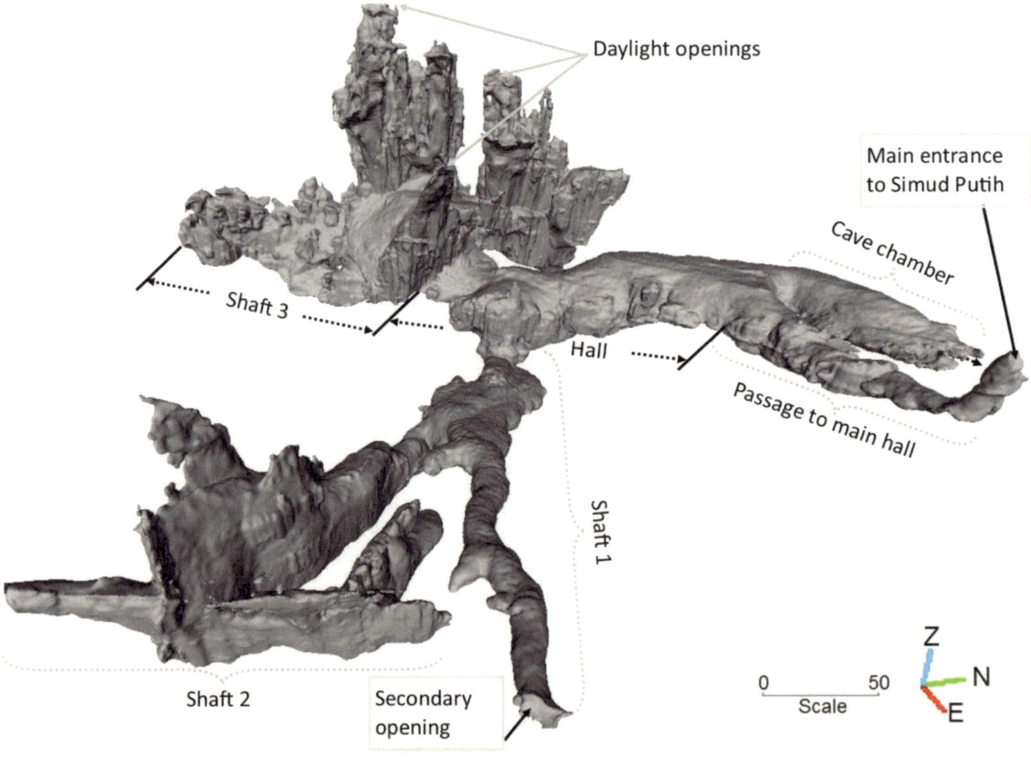

Figure 3. 3D model of the upper level cave (Simud Putih) produced in MeshLab with ≈ 28.3 million points at an octree depth of 10.

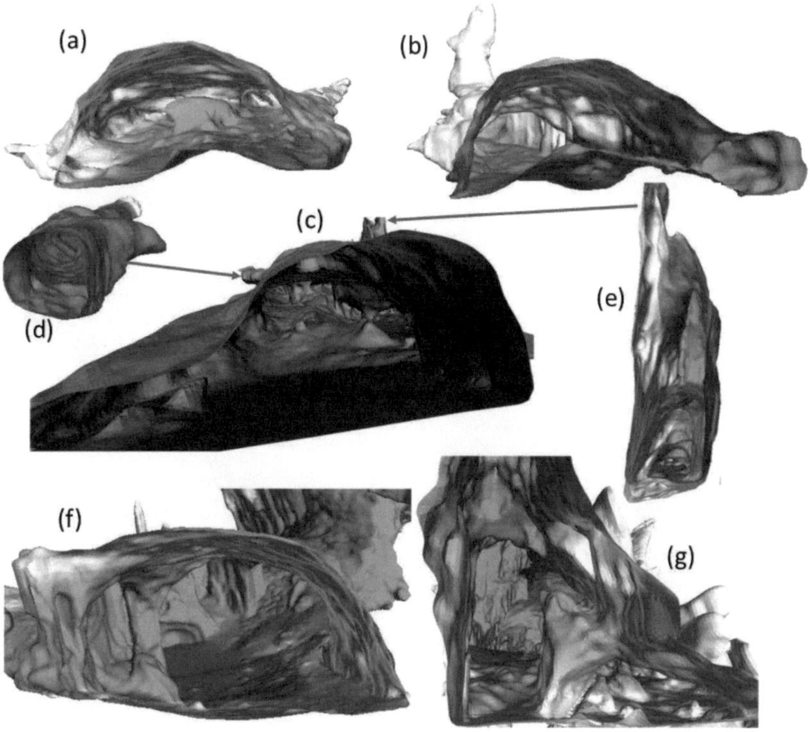

Figure 4. Graphic description of the cave cavities: (a) main chamber viewed W-E, (b) the horn-shaped entry passage viewed W-E, (c) secondary chamber where shaft 1 (d) and shaft 2 (e) are connected, (f) main hall viewed from N-E to S-W at the centre of the hall and (g) shaft 3 viewed towards the north direction.

characteristics of the karstic landform. Precisely, the model shows that the passages mostly still retain their original phreatic tube except for shaft 3 and the bourgeoning section of shaft 2. Although there is no active stream flowing in it, there is evidence of flooding during heavy precipitation particularly within the secondary chamber that connects shafts 1 and 2 to the terminal section of the hall which forms the lowest elevation of the cave. The floor also appears more solidly intact except with some isolated rock piles but is a relatively gentle undulating surface of interchanging concave and convex arc that cambers to each side creating shallow erosional bench at the edges (**Figure 4a–4c** and **4f**). **Figure 5e** and **5g** depicts roof patterns that may have originated from lateral strike slip and fault-controlled attritions. The wall largely comprises tri-oval conduit arm, buttress, vertical and semi-circular walls holding the dome roof.

4.2. Cave morphometry

An overview of the shape characteristics of each segment is shown in **Figure 5**. The sectional drawing improves the understanding of the internal outline of conduits at the roof and floor sections. It can be seen that the cave passages vary in height and width, which implies that the cavities either respond differently to denudation or are developed across different timelines. The shapes of the passages vary from simple flute morphologies to rather complex features with overhang, pillars, horn-like extension and step-tile cathedral interior formations (**Figure 5**).

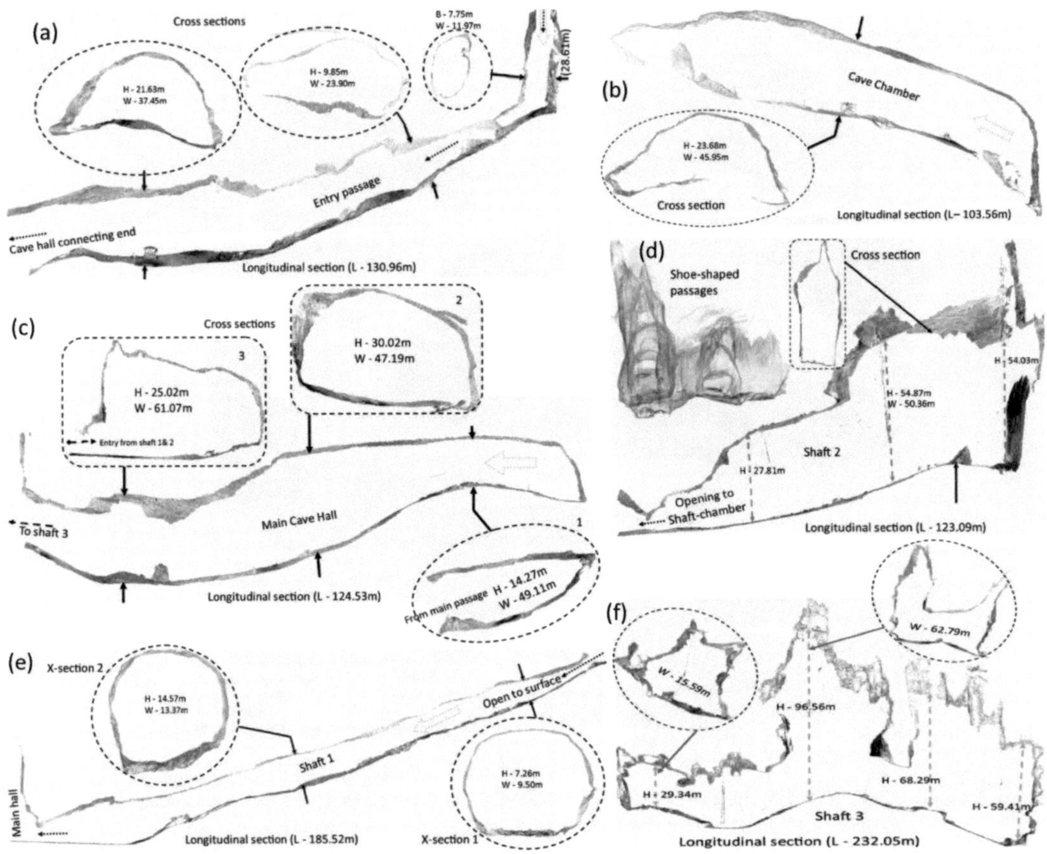

Figure 5. Side and cross-sectional view highlighting passage outline of (a) entry channel, (b) chamber, (c) main hall, (d) shaft 2, (e) shaft 1 and (f) shaft 3.

Dimensionally, the cave extends linearly from sub-metre openings to impressive passage width of up to 65 m, varying heights and several metres' penetration depth developed along major tectonic features like faults, fold cores and bedding contact. All the passages descend to the more distant part to converge at a central area down south of the main hall that forms the lowest part of the cave network. The floor at this minimum elevation is relatively flat (**Figure 5c** and **5e**) and filled with sediments, pebbles and flowstone. The impact and degree of weathering activities are more pronounced at the cave ceiling with exciting irregular shapes that provoke closer examination particularly in shafts 2 and 3. In general, the mesh model facilitates quantitative measure with better accuracy (**Table 2**).

The entrance to the cave is located in vertical rock walls of about 29-m depth from the top of hill and continues deeper along a steep rock wall down the hillslope to the main hall (**Figure 5a**). The cavity becomes wider as it advances towards the hall, forming a horn-shaped hollow and terrace outline at the roof. Entry passage and chamber connect the large lofty hall that constitutes the primary trunk of the cave system. The chamber (**Figure 5b**) has a high-domed formation that looks like the hard-upper shell of tortoise (carapace) but at the base, a

Total length (m)	**1,043.05**
Total floor area (m^2)	38,575.00
Passage surface area (m^2)	155,414.00
Passage volume (m^3)	735,226.00

Table 2. Summary of basic geometric properties of the model.

convex silhouette slopes to the foot of the wall at both sides. The hall has a rectangular shape of about 50–70-m width, a buggy-top roof of up to 30-m height and a basin morphology landform at the floor (**Figure 5c**).

Passage layout of shaft 1 (**Figure 5e**) is unique; the conduit forms a long tubular flute-shape passage with oval outline in cross section along the entire length of the strait. Like the entry passage, the conduit has erosional contact with the surface but at a relatively gentle gradient that allows a steady slipping down of rainwater and eroded materials along the slope. It is probably the youngest passage because it exhibits undisturbed early stage phreatic tube formation with scallops freshly sculptured in the roof and wall of the tube.

In contrast, shafts 2 and 3 show completely different morphologies especially at the roof. The former is a boot-shaped twine passage lying 30-m apart and is linked by a narrow vertical channel of approximately 42-m height (**Figure 5d**). The right passage which is about 60-m long terminates at a dead end while the other "leg" extends to about 123 m and joins a small chamber through a narrow opening that connects it with shaft 1. The two have similar morphologies that take the shape of a wellington boot. Shaft 3 (**Figure 5f**) produces a typical cathedral morphology with step-tiles roof. The interior has been severely weathered, thus creating different karstic formations including enormous cavern resulting from breakdown chamber, multi-storey passages, avens and epikarstic seepage.

As a conclusion, the mesh model provides adequate information to describe the general cave network, internally and externally, which has never been done. Furthermore, we were able to produce the cross section and obtain accurate morphometrics of the model (**Table 2**). However, the coarse nature of the source point cloud and lack of texture information limit the amount of details that can be extracted from the model.

4.3. Identification of micro-morphological features

Intensity-based texture model shows all the fine geometric details that bring true-to-life virtual caving with complete representation of the interior "infrastructure" (**Figure 6**). In contrast to the mesh model, the texture model reveals all the signs of epigenetic speleogenesis in addition to identification of fine detailed morphological features. Apart from these, dimensional measurements of sub-millimetre accuracy can be obtained. Certainly, the challenge of visualizing cave interiors due to lack of illumination is solved with this approach.

Generally, it can be noticed from the texture model that all the passages, except shafts 2 and 3, have scallops on their wall and roof, indicating the extent to which the original phreatic

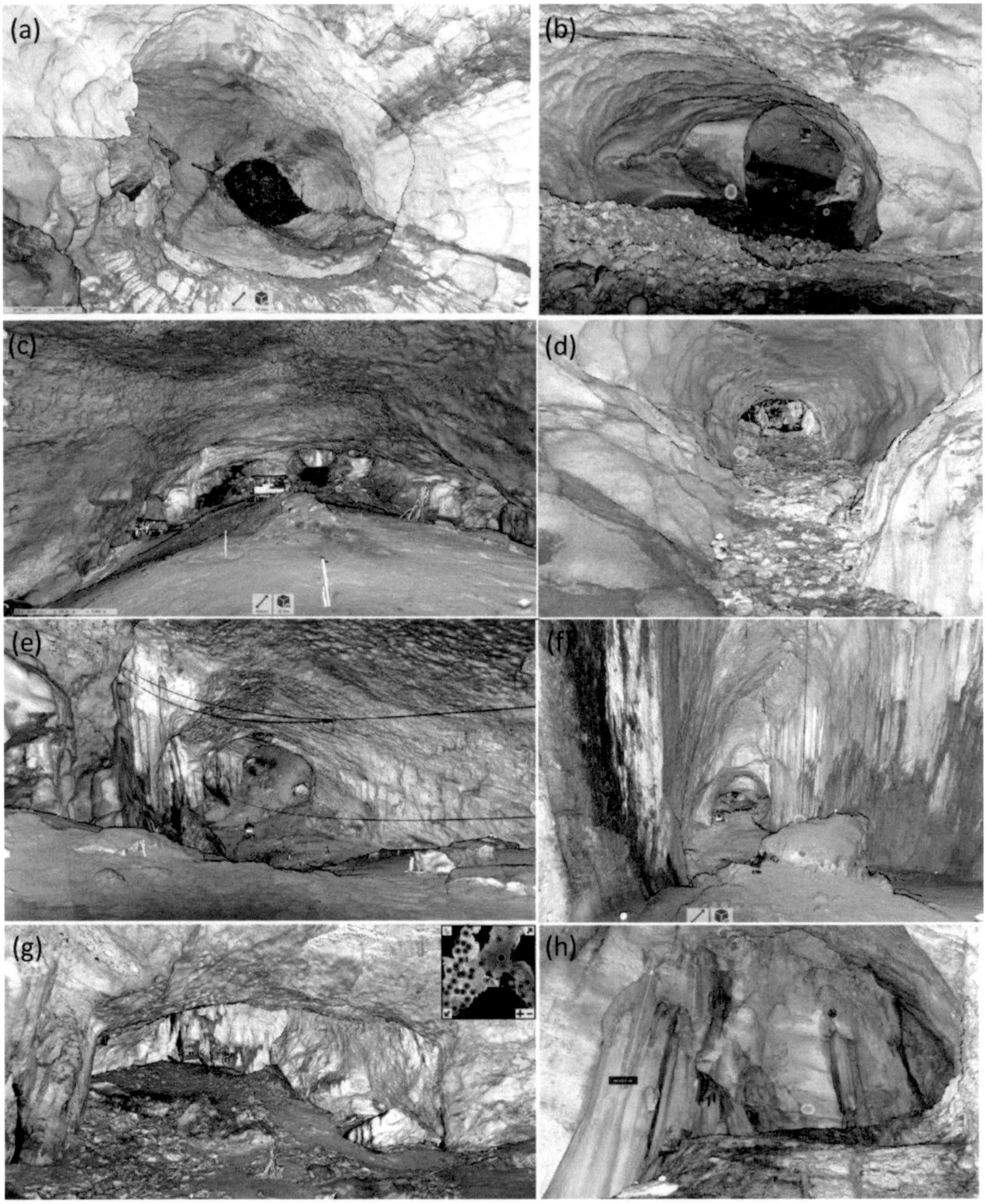

Figure 6. Intensity-based texture model of Simud Putih—topmost plate shows (a) the main entrance viewed from the base of the collapse sinkhole and (b) E-W direction towards the hall; plate (c) is the chamber viewed W-E, (d) shaft 1 viewing the entrance from inside, (e) main hall viewed to the southern end, (f) shaft 2 viewed northward; plate (g) views shaft 3 from the hall where the passage opens to it while (h) shows a section within shaft 3.

tube is preserved. The rate of passage modification seems to be at slow pace in the older passages. For example, within the chamber and main hall, scar of recent deformation can be found. Tonner variation distinguishes recently developed surfaces from the older ones. It can be observed that the inner section of the entry passage, main hall, chamber and shaft 3 cavern belongs to the old formations. The old formations are discernible with much stable appearance, particularly at the roof which appears to form the base of harder rock strata. Those parts of the cave are favourable roosting locations for swiftlet birds in large numbers. Interestingly, bats and swiftlet birds can be differentiated from the texture model.

It is very clear that the development of the cave started from inside and, over time, eroded to the surface to provide entrances (see **Figure 6a** and **6d**). Similar imminent contact with surface is observed in shaft 3 which currently provides daylight openings along fractured zones that extend to about 118 m in length at the roof. The fault joins series of rock discontinuities (fault and fracture on the wall) that extend to the cave floor (see **Figure 7a** and **7c**). The roof around the daylight opening may eventually become collapse doline.

The floor of the cave is made up of fine mud sediments, rock debris, rock piles, plastic bottles, dendritic sediments, plates and polythene bags washed into the cave or abandoned by cavers. In the more weathered passages, a pile of rocks that broke away from the cave roof can be found on the floor. A stream of pebbles can be seen at the base of the sinkhole that form the entrance and also along shaft 1 passage which are transported into the cave and deposited along the passage floor. The occurrence of the cave passages and the pattern of post-speleogenetic decay have been largely influenced by the deformation characteristics of the rock rather than biogenic modifications because no perforated cave walls are observed in Simud Hitam.

Apart from shaft 3, the cave walls are solidly intact (**Figure 6**). Few base undercuts and wall notches are present at lower end of each passage where the eroded rocks have direct contact with the wall. Further, vestiges of fluted bay can be seen in shaft 3 though not as pronounced as those found in Simud Hitam. Other interesting wall morphologies include semi-circular pillar, cliff and hanging wall (**Figure 7b** and **7d**). A number of semi-circular pillars of varying diameters were identified on the wall of shaft 3. These pillars are made up of accumulated calcites dissolved in dripping waters from the cave roof. Evidence of this can be seen in the evolving shining glossy formations like stalactite (STT) and stalagmite (STM) approaching each other to form an unbroken pillar. All the pillars extend from the roof to the cave floor, measuring up to 20 m in height. We also identified other features such as cliffs and hanging walls (**Figure 7b**) within shafts 2 and 3. These features are certainly remnants of collapse blocks that are resistant weathering.

Generally, no lace-like or overtly perforated ceiling surface was observed; swiftlets take advantage of rock discontinuities within the more secured sections of the cave as niches to roost. The common morphological features identified in the cave roof include scallops, bell holes, ceiling pockets, roof avens, ceiling flush dome, seepage and hanging rocks (**Figure 8**). Except for the last three, the origin and development of these features have been well discussed in the literature [29, 30]. Scallops are predominantly retained in the more recently developed younger surfaces while traces of it are apparent in the older hard rock surfaces, but roof avens are common to weak rock surfaces. On the other hand, ceiling and conch pockets and bell holes are conventional outcomes of passive weathering on the older surfaces (chamber, hall, shaft 3 corridor and entry passages).

Ceiling flush dome is similar to semi-spherical flush mounted ceiling light, but unlike the bell-hole, the base flushes with the cave roof measuring between 20 and 80 cm in diameter,

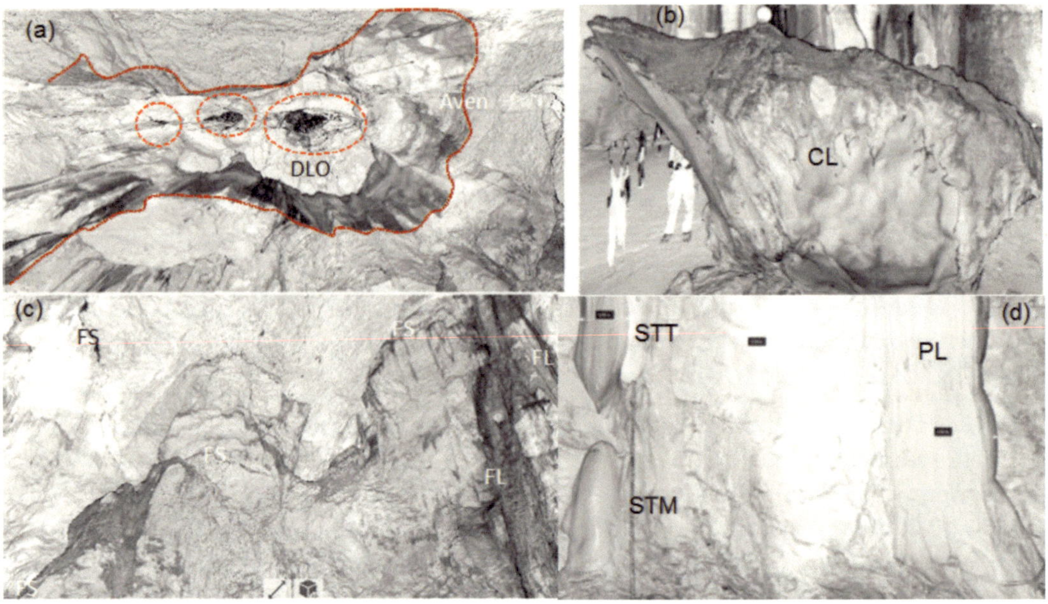

Figure 7. Rock surfaces: (a) daylight opening in the roof, (b) cliff, (c) fault and fractured wall and (d) semi-circular wall pillar (PL), including stalactite and stalagmite.

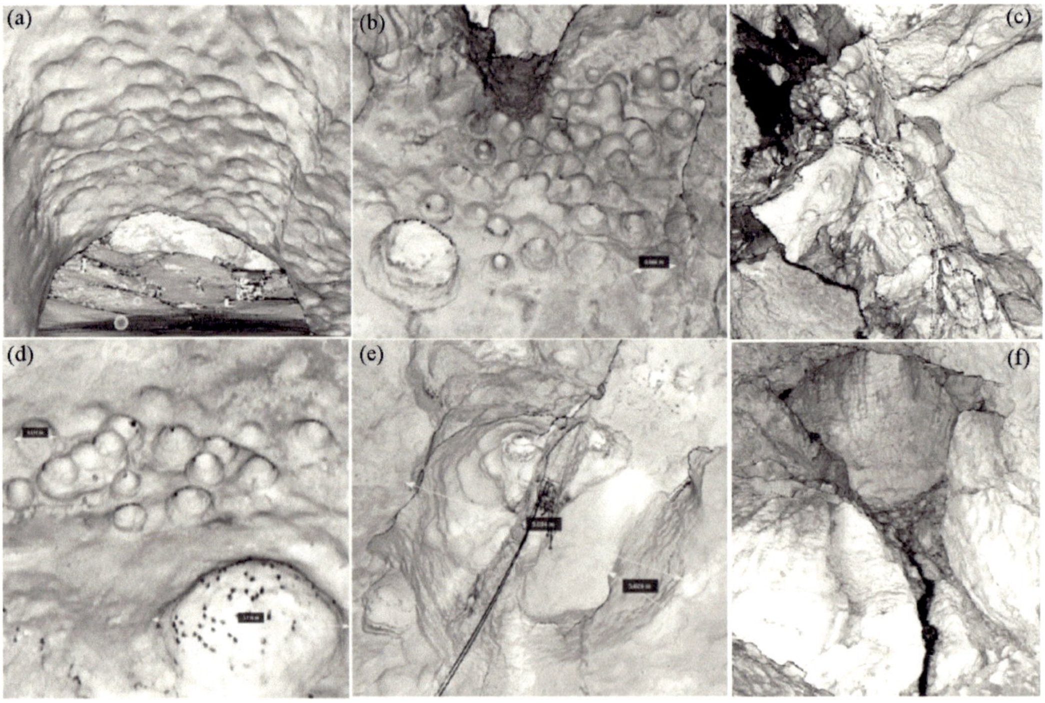

Figure 8. Cave roof morphologies: (a) scallops, (b) ceiling flush dome, (c) hanging rocks and fractures, (d) ceiling pockets and bell-hole, (e) ceiling aven of about 9 m in diameter and 5 m deep into the roof and (f) seepage.

suspended 30 cm from it. This particular roof feature is found in shaft 3. The feature may have originated from accumulation of calcite dissolved by water infiltration through carbonate rock. Another feature identified on the roof is the mass of rocks hanging on the roof, we call this hanging rock. This feature exists within unfavourable discontinuities especially where such fractures form follicular dendritic network that partitions weak rocks into loosely fragmented components. If the fracture reaches the surface, water may penetrate into the cave to create seepage. Seepage is a catalyst to cave deformation because it weakens the rock and speeds up the rate of weathering. A number of these features were identified in the roof and wall of the cave at different locations. Identification of seepage is important because it can help in predicting direction of future passage development and potential inlets.

4.4. Cave art visualization

Lack of efficient tools to combine geometric data and images has long made the process of producing virtual reality models from 3D scanning and texture information a difficult exercise. But with the emergence of new sensors and processing tools, the time and efforts needed to generate this product have been efficiently optimized, and the result obtained is so accurate that objects of sub-millimetre dimension can be detected and measured. As evidenced from virtual tour of the cave, people have been registering their presence in the cave by penning down ideas in written inscriptions and rock carvings over the years by striking directly on the rock surface (**Figure 9**).

One of the significances of the study is the demonstration of the ability of the high-resolution 3D texture model to detect carved rock and paintings on the wall and roof of the cave at sub-millimetre resolution. This result is novel and provides potential economic opportunities through tourism if well harnessed. A collection of archaeological treasures including pictographs, petroglyphs and inscriptions is found on the walls of the cave passages. The age of these engravings is not known, and no record is given about them. So, we believe they have no link with medieval traditional, spiritual or cultural use. It can be seen that some of the carvings are undergoing surface erosion and deterioration, and such can be quantitatively measured from the digital model. Identified litters in the floor of the cave are also bottles, plastics and disposable serving plates that are products of modern-day industries.

Exploiting the full potentials of this approach may play a significant role in geo-archaeological studies in the near future. Particularly, with the increasing availability of new high-quality digital sensors for data collection and efficient tools for data processing and analysis, qualitative recording, documentation and assessment of cave art are within the reach of archaeologists. The principal limiting factors to the adoption of this approach are the costs of acquiring the equipment and the considerable time and efforts required to train the users. While the added value to 3D documentation might not justify the investment, the multifunctional value of the data collected is far-reaching. A case in point is the increasing multidisciplinary perspective of cave research that has brought closer cooperation among archaeologists, geomatic engineers, geologists and other non-technical professionals [2]. This type of collaboration is required to offset cost implications through shared responsibilities without compromising precise documentation that yields better results.

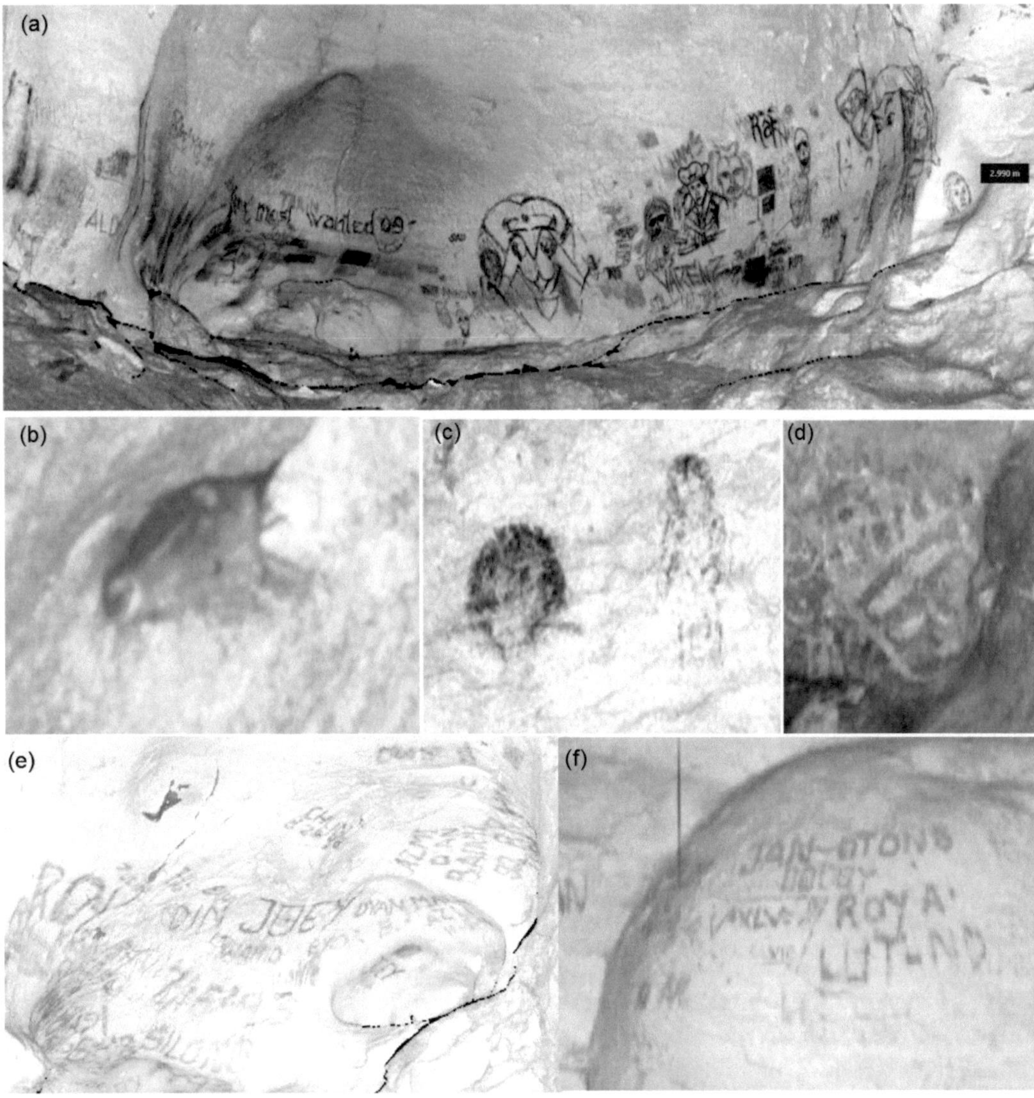

Figure 9. Cave arts on the wall—(a) paintings in shaft 1 (b, c), a picture of a goat head, and painting in shaft 3 and (d) cave engravings in the chamber and (e, f) cave inscriptions in entry passage.

5. Conclusion

Experience from this experimental work with intensity-based 3D texture model indicates a promising future for refined exploration of cave interiors at exceptionally high level of accuracy and details in areas that would challenge even the most adventurous caver. It is a paradigm shift from the era in which darkness in cave was seen as a major obstacle to visualization and information extraction. Given the versatility of laser-scanning data as demonstrated in this chapter, it is plausible to submit that further research will be required to adequately harness the invaluable information therein embedded. This we believe is necessary to enhance further cave research and development. It is expected that as processing capabilities improve along with powerful sensor development, more applications are expected to emerge. The design of powerful modelling packages like ReCap 360 which performs parallel computing for 3D modelling and texturing bypasses the usually time-consuming matching of laser scans and photos and the associated manual editing. Similarly, increasing availability of other third-party point cloud oriented application software (open source and commercial) is broadening the sphere of possibilities with laser-scanning data. This development would change the phase of cave investigation across the different professional interest groups.

In terms of details, the full-scan model offers incredible clarity; objects as small as giant spiders and millipedes can be detected and measured. This is why it was evaluated for archaeological documentation. In the future, transmission of this information on the web can multiply its values and enhance knowledge sharing across platforms and agencies. The limitation of this approach is that it has no spatial analysis capability. Again, the problem of poor illumination still persists with natural RGB colour images, although it has little to do with geomorphological and geological applications such as those presented in this chapter.

Acknowledgements

This research is supported by Ministry of Higher Education, Malaysia research grant (FRGS/1-2014- STWN06/UPM//02/1) with vote number 5524502 and University Putra Malaysia research grant (GP- 1/2014/943200). The authors like to thank Professor Dr Manfred Buchroithner for joint terrestrial laser-scanning expedition funded by the National Geographic Society.

Author details

Mohammed Oludare Idrees and Biswajeet Pradhan*

*Address all correspondence to: biswajeet24@gmail.com

Department of Civil Engineering, Geospatial Information Science Research Center (GISRC), Faculty of Engineering, Universiti Putra Malaysia, Selangor Darul Ehsan, Malaysia

References

[1] Tolan-Smite C. Human occupation of caves. In: Encyclopedia of Caves and Karst Science. Taylor and Francis Group, New York; 2004. pp. 919-924

[2] Burens A, Grussenmeyer P, Guillemin S, Carozza L, Lévêque F, Mathé V. Methodological developments in 3D scanning and modelling of archaeological French heritage site: The Bronze Age painted cave of "Les Fraux", Dordogne (France). International Archives

of the Photogrammetry, Remote Sensing and Spatial Information Sciences. 2013;**XL-5/W2**(131-135):2-6

[3] Price L. Species diversity and food-web complexity in the caves of Malaysia. Ambient Science. 2014;**1**(2):1-8

[4] Idrees MO, Pradhan B, Buchroithner MF, Shafri HZM, Bejo SK. Assessing the transferability of a hybrid Taguchi-objective function method to optimize image segmentation for detecting and counting cave roosting birds using terrestrial laser scanning data. Journal of Applied Remote Sensing. 2016;**10**(3):1-16

[5] Gallay M, Hochmuth Z, Kaňuk J, Hofierka J. Geomorphometric analysis of cave ceiling channels mapped with 3D terrestrial laser scanning. Hydrology and Earth System Sciences. 2016;**2016**(February):1-48

[6] Idrees MO, Pradhan B. A decade of modern cave surveying with terrestrial laser scanning: A review of sensors, method and application development. International Journal of Speleology. 2016;**45**(1):71-88

[7] Lundberg J, McFarlane DA. Post-speleogenetic biogenic modification of Gomantong Caves, Sabah, Borneo. Geomorphology. 2012;**157-158**:153-168

[8] Hutchison ES. Geology of North West Borneo: Sarawak, Brunei and Saba. Elsevier, Amsterdam; Boston 2005

[9] Noad J. The Gomantong Limestone of Eastern Borneo: A sedimentological comparison with the near-contemporaneous Luconia Province. Palaeo. 2001;**175**(2001):273-302

[10] Mcfarlane DA, Buchroithner M, Lundberg J, Petters C, Roberts W, Van Rentergen G. Integrated three-dimensional laser scanning and autonomous drone surface-photogrammetry at Gomantong Caves, Sabah, Malaysia. In: 2013 ICS Proceedings: Karst and Cave Survey, Mapping and Data Processing—Poster, Czech Speleological Society, Czech Republic, Brno 2013. pp. 317-319

[11] Kingston T. Research priorities for bat conservation in Southeast Asia: A consensus approach. Biodiversity and Conservation. 2010;**19**(2):471-484

[12] Abdullah MT, Paul IV, Hall L. A frightful stairway to cave bats in Borneo. Bat Conservation International. 2005;**23**(3):11-13

[13] Lim KH, Khoo CK, Laurentius NA, Yeo BK. A preliminary report on the surveillance of Highly Pathogenic Avian Influenza (H5N1) and Newcastle Disease (ND) viruses in edible bird nest swiftlet (*Aerodramus fuciphagus* and *Aerodramus maximus*). Malaysian Journal of Veterinary Research. 2012;**3**(1):1-5

[14] McFarlane D, Roberts W, Buchroithner M, Van Rentergem G, Lundberg J, Hautz S. Terrestrial LiDAR-based automated counting of swiftlet nests in the caves of Gomantong, Sabah, Borneo. International Journal of Speleology. 2015;**44**(2):55-60

[15] Dublyansky Y, Roncat A, Spötl C, Dorninher P. Hypogene cave morphology at high resolution: Full 3-D survey of Märchenhöhle (Austria). In: Chavez T, Reehling P, editors. Proceeding of Deepkarst 2016: Origins, Resources, and Management of Hypogene Karst; 11-14 April; Carlsbad, New Mexico, USA: National Cave and Karst Research Institute; 2016. pp. 183-187

[16] Chandelier L, Roche F. Terrestrial laser scanning for paleontologists: The Tautavel Cave. In: XXII CIPA Symposium: Digital Documentation, Interpretation & Presentation of Cultural Heritage; 11-15 October 2009; Kyoto, Japan. ICOMOS, Paris, France. 2009. pp. 1-5

[17] Tsakiri M, Sigizis K, Billiris H, Dogouris S. 3D laser scanning for the documentation of cave environments. In: 11th ACUUS Conference: Underground Space, Expanding the Frontiers; 10-13 September 2007; Athens, Greece. ACUUS, Greece. 2007. pp. 403-408

[18] Besl P, McKay N. A method for registration of 3-D shapes. IEEE Transactions on Pattern Analysis and Machine Intelligence. 1992;**14**(2):239-256

[19] Huber D. The ASTM E57 file format for 3D imaging data exchange. In: Proceedings of the SPIE Vol. 7864A, Electron Imaging Science and Technology Conference (IS&T), 3D Imaging Metrology, SPIE, USA. 2011. pp. 1-9

[20] CloudCompareV2.8. GPL Software [Internet]. 2016. Available from: http://www.cloud-compare.org/. [Accessed: 15 August 2016]

[21] Cignoni P, Ranzuglia G. MeshLab. Visual Computing Lab [Internet]. ISTI–CNR. 2014. Available from: http://meshlab.sourceforge.net/. [Accessed: 26 December 2016]

[22] Silvestre I, Rodrigues JI, Figueiredo MJG, Veiga-Pires C. Cave chamber data modeling and 3D web visualization. In: Proceeding of International Conference Information Visualisation, IEEE Computer Society, USA. 2013. pp. 468-473

[23] Silvestre I, Figueiredo M, Veiga-pires C. High-resolution digital 3D models of Algar do Penico Chamber: Limitations, challenges, and potential. International Journal of Speleology. 2015;**44**(January):25-35

[24] Lerma JL, Navarro S, Cabrelles M, Villaverde V. Terrestrial laser scanning and close range photogrammetry for 3D archaeological documentation: The Upper Palaeolithic Cave of Parpalló as a case study. Journal of Archaeological Science. 2010;**37**:499-507

[25] González-Aguilera D, Muñoz-Nieto A, Gómez-Lahoz J, Herrero-Pascual J, Gutierrez-Alonso G. 3D digital surveying and modelling of cave geometry: Application to Paleolithic rock art. Sensors. 2009;**9**(1):1108-1127

[26] Gonzalez-Aguilera D, Rodriguez-Gonzalvez P, Mancera-Taboada J, Muñoz-Nieto A, Herrero-Pascual J, Gomez-Lahoz J. Application of non-destructive techniques to the recording and modelling of Palaeolithic rock art. In: Wang C-C, editor. Laser Scanning, Theory and Applications. InTech; Open Science Publisher. 2011. pp. 305-326

[27] El-Hakim SF, Fryer J, Picard M. Modeling and visualization of Aboriginal rock art in the Baiame Cave. In: Proceeding of XXth Congress of the International Society for Photogrammetry and Remote Sensing; 12-23 July 2004; Istanbul, Turkey: Comm V, Work Gr V/2; 2004. pp. 990-995

[28] Autodesk. Autodesk Software for Students, Educators, and Educational Institutions [Internet]. Autodesk Education Community. 2017. Available from: http://www.autodesk.com/education/free-software/featured. [Accessed: 4 November 2016]

[29] Lacanette D, Large D, Ferrier C, Aujoulat N, Bastian F, Denis A, et al. A laboratory cave for the study of wall degradation in rock art caves: An implementation in the Vezere area. Journal of Archaeological Science. 2013;**40**(2):894-903

[30] Plan L, Tschegg C, De Waele J, Spötl C. Corrosion morphology and cave wall alteration in an Alpine sulfuric acid cave (Kraushöhle, Austria). Geomorphology. 2012;**169-170**:45-54

Chapter 2

Aspects of Cave Data Use in a GIS

Gáspár Albert

Additional information is available at the end of the chapter

http://dx.doi.org/10.5772/intechopen.68833

Abstract

This paper gives an overview about the theoretical and technical aspects of a geographic information system (GIS), which can provide a framework for scientific (hydrological, morphological, geological, etc.) analysis of cave survey data. It is emphasized that a GIS containing archive cave data is important because the information often is irreplaceable (the cave environment has changed), or the resurveying may harm the cave. Thus, it is proposed that a GIS of cave data be multidisciplinary to avoid unnecessary resurveying of caves. To produce such a system, one has to bear in mind many aspects, which is not always evident for the practicing scientists. Cave surveys produce spatial data, either it was measured with measure tape and compass or with a LiDAR station. It is a major issue in cave data processing that the spatial data produced in various surveys do not fit together due to the different methods and coordinate systems, or because of the various data types, which make it hard to syllabize the similarities between different sets of data. The paper focus on how to work with archive and new survey data, and how to handle maps, scans, and sampling data in one information system. From the aspect of data transfer, three main functionalities of a GIS are distinguished: *processing*, *storing*, and *representation* of the information. Discussing the theoretical and practical backgrounds of these functionalities, the paper presents the best practices of building a GIS from archive and newly measured data, emphasizing the importance of procedures like data management, quality control, and automation. The paper shed lights to the various data types that are usually related to cave surveys, to help cave scientist to control the data management and understand (and apply) the automatisms. Also, the probable technical parameters of future cave surveillance systems are discussed.

Keywords: cave surveying, GIS, archive data, cave data management, GIScience

1. Introduction

Cave survey data is collected for several reasons. Cave maps, 3D models, and parametric data for many disciplines of science can be created and extracted from the raw measurements. The amount of the collected data during a cave survey is much larger today than it was decades ago, thanks to the evolution of surveying tools and methods [1]. Such amount of data is created and handled in several ways depending on the skills and intents of the surveyor/processing staff or the scientists involved, and sometimes due to the lack of geographic information system (GIS) experience, the aims cannot be fully achieved [2]. However, the resolution of the collected information does not always match with the aims of the project, and this also may cause improper conclusions in a scientific project or inaccurate cave maps [3, 4]. In the classic times of speleological surveys—using measure tape and compass—the collected data was rather insufficient for scientific purpose and now with the use of terrestrial laser scanning (TLS) stations, it is rather too numerous to efficiently work with for many users [5].

Methodological papers about the analysis of cave data concentrate usually on the theoretical and not the practical aspects of data processing. For example, in the case of a new cave survey, the comparison of the newly collected data and the archive data is reported in several studies [3, 4], but the details of how the different data packages were incorporated in one system remain in the background. For several spelunkers—however—these details are possible sources for errors due to the lack of experience with coordinate transformations or management of data files, and programs, which are involved in this process. Such comparisons are often used to illustrate the higher accuracy of the new surveying methods, and while this is reasonable, the archive data often preserves notes and observations, which bear scientific importance (e.g., locations of bat colonies, archeological specimens, water sampling, etc.). Moreover, archive data often preserved cave conditions that have changed since the time of the survey. In an optimally built GIS, both the new and the archive data are positioned correctly and the database of the surveys can be queried simultaneously.

The aim of this paper is to highlight the cave data processing from the aspect of a geographic information system (GIS), and to demonstrate that such information system can be used to help scientific projects to combine newly measured and archived data. Using a GIS means using database tables, data transformation tools, statistical and spatial analysis, filtering and extracting certain parameters from the raw data, and finally (but not necessarily) visualizing the results in 2D or 3D. To do this efficiently, one must possess knowledge about such processes, and the know-how is so complex that it has become an advancing discipline: the GIScience.

2. The nature of cave survey data

All cave survey project starts with a plan to do something for a certain purpose. At this point, the surveyors' knowledge and the available instruments will determine the quality of their future data. Although, contemporary surveyors have already upgraded their instruments, still there are lots of data from the predigital times. Most of the caves in the world were surveyed (at least partially) with measure tape and compass, by speleologists progressing in the cave passages from station to target points [6]. The data consists of several records containing an array of distance, dip, and azimuth measurements making a 2–10 m long spatial vector from each record. The sequentially joining vectors form a 3D network (**Figure 1**), and each

http://dx.doi.org/10.5772/intechopen.68833

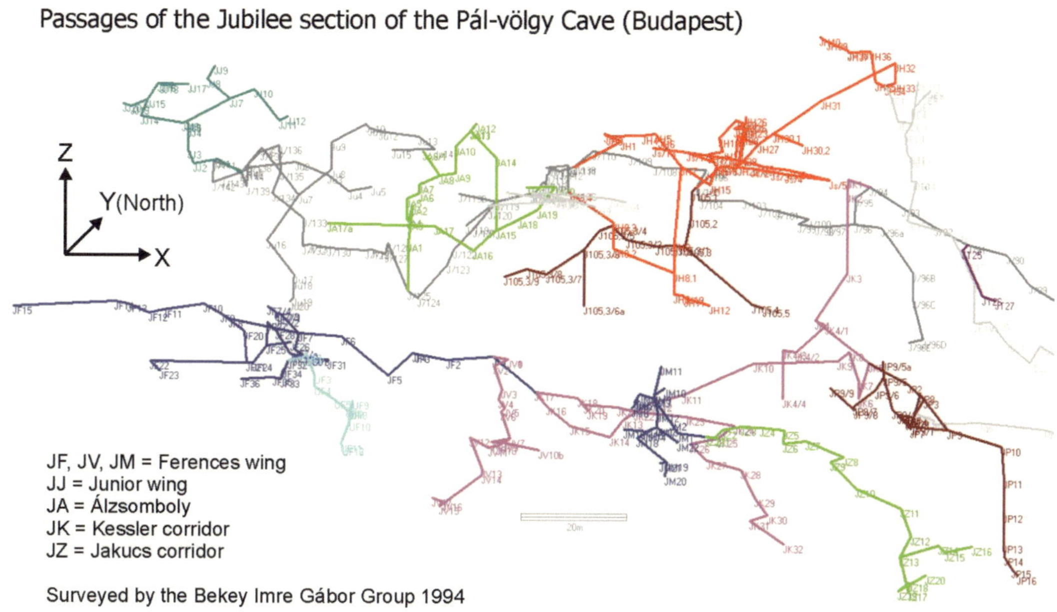

Figure 1. Typical line plot of a survey network from the Pál-völgy Cave (Budapest). The shades (colors) indicate different surveys in various times. Alphanumeric characters indicate the identification codes of the measured stations.

node of this network can be represented with x, y, z coordinates. Although, the survey produces its own coordinate system, where the origin is the survey base point (entrance station), the whole set can be inserted in real geographic position defined in a Euclidian geodetic reference system (X, Y, Z) if the cave entrance was measured with geodetic instruments (i.e., total station or GPS). Additionally, the width and height of the cave passage can also be measured at each station to provide data for 3D cave models [7, 8]. The punctuality of the measurements can be enhanced if the network is measured backwards too (back to the original station), or if loops are present in the cave where the closing point of the loop is identical to one of the previously measured stations. This latter measuring sequence in the loop may also be measured reversely, and the average error can be distributed among the stations of the loop [3, 6]. This method (*loop closure*), however, was not always followed by the surveyors in the past due to several reasons. Most often, the backward measurements are missing, and the whole passage system was probably measured in different times by different people, which means that the condition of the data is varying. Usually, the most problematic part of creating a consistent GIS is the harmonization of several unit systems, data structures, content types, and coordinate systems.

3. Inputs and outputs

The condition of the created survey data will determine how much time is necessary to build a working GIS. The GIS is always built to serve as a "tool" to achieve the aims of a certain project (even if onward use is also in the hat), so the planning should consider the possible sources of the cave survey data. Basically three scenarios are the most frequents: (1) the data is to be salvaged from archives, (2) new survey is done, and (3) partly new and partly archive data are processed. In all cases the created data will be *processed*, *stored*, and *visually represented* in some way.

- The data *processing* tool is responsible for the digital recording of text (notes, reports), alphanumeric and logical (true/false) information, images (photos, scanned documents), and vector geometry (line plot maps, sections).
- Data storage devices provide secure data *store* and availability of alphanumeric and logical data, images, and texts in digital or digitized documents. Digital data is located on mass storage devices in its appropriate format and accessed via database and file management tools.
- The tool, which *represents* the data as a 2D map or a 3D model is a complex application, which not only visualize the data, but most often serve as the GIS environment. It makes possible to access not only the visual representation of the data, but all the collected information from the database.

These main functionalities are accessed through programs directly developed for cave survey data management (Compass, Visual Topo, TopoRobot, and Therion), or with well-known applications developed for general data management (Excel, ArcGIS, AutoCAD, and many other). In either way, the application itself becomes organic part of the information system of the survey project with all of its data formats and processing algorithms. In some cases, the applications themselves are only applying external programs or scripts [9], increasing the complexity of the whole system. Selecting the tools for these functionalities is one of the most crucial parts of planning of the GIS. The aspects should include the consideration of the aim of the project firstly, but later the usability and the spatial dimensions should also be tackled. The more we rely on archive data, the more we should involve noncave mapping applications to bring the data into acceptable form.

3.1. Data types of an archive survey

Using only archive data is appropriate in the case when the project cannot afford new surveys either because of time or financial limits, or the cave has an endangered environment where even the survey may cause serious damages. The spatial resolution of the data is usually suitable for morphometric analysis [10, 11], but realistic 3D models cannot be created. *Archive survey* data can be collected from paper-based documentations (reports, notebooks, maps, and published papers). In this case, the GIS will be composed of the digitalized forms of these documents, and will always incorporate the following components: database, digital map, original documents (scanned), and the tools (programs). In a general purpose GIS, the main characteristics of these components are the followings:

1. The *database* is to be built from the notebook records. Although, the temptation is usually great to skip seemingly the irrelevant information (e.g., the condition of measuring) during digitization, it is always useful to fill the database with attributes like "CONDITION", or at

least put such remarks in a "NOTE". The database usually contains the following attributes: date of survey, station-id, target-id, distance between station and target, dip, azimuth (angle from the north in clockwise rotation), width and height of the passage at the station. If the entrance points are measured with GPS, the database can be completed with the latitude, longitude, and elevation of each point using Euclidian geometry and the vector data.

2. The *digital map* is created from scanned paper maps usually to provide additional information about the morphology of the cave. Maps contain the outlines of the passage levels indicating the characteristic morphology of the walls and the main artifacts. Additionally point-like objects, names, transversal, and longitudinal sections are also displayed (**Figure 2**). The maps which are suitable for morphological analysis have usually large scale (>1:1000), but rarely contain geographic or geodesic coordinates, so the first step is the "georeferencing", which means that identifiable points (e.g., the marked stations) on the map are associated with their geographic coordinates. These coordinates can be calculated from the base point. The map processing starts with the determination of the data types, which are selected to be digitized. Each map data types (points, texts, lines, polygons) will be stored as a graphical element associated in one or more files. The processing tool will determine if the created

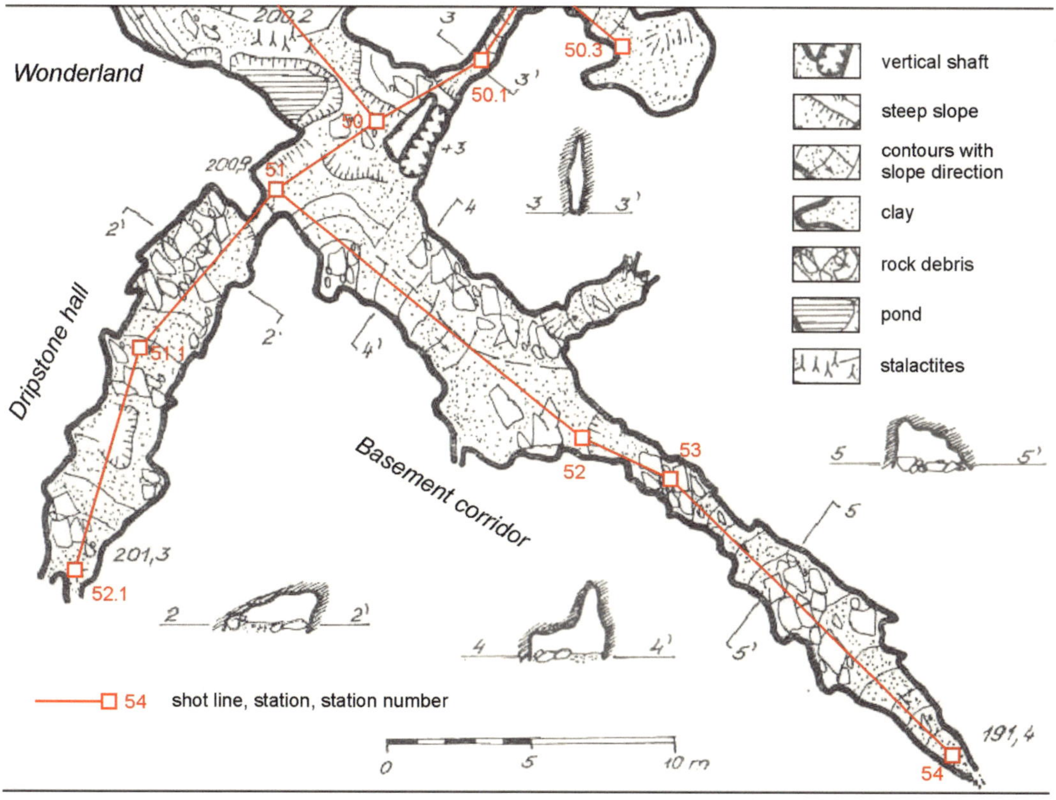

Figure 2. Part of a typical cave map indicating transversal sections. The map [12] shows a small part of the Pál-völgy Cave System (Budapest).

files are suitable for GIS, thus, it is most appropriate to use a GIS program directly (e.g., ArcGIS, MAPINFO, or QGIS). These programs are also suitable to do the "georeferencing" with the help of the previously processed *database*.

3. The *digital archiving* of the original data is advised in the case of the notebooks and necessary in the case of the maps. The scanned documents are usually in pdf or jpg format, and obviously stored on a hard drive. However, the location of the archived data is highly relevant from the aspect of the GIS, because the map processing tools will record the name and the source (folder) of the map file during the digitization process, so it will cause problems if the folder structure or the file name is changed after the process has started. This is also true for the created files at the end of the digitization: we will face with file access problems if their location or name is changed.

4. The *preferred tool* can be one of the cave surveying programs, but depending on the condition of the recorded measurements other processing tools can also be appropriate. If the records are accurate (all the previously listed attributes are present), the surveying program can produce a representation too and calculates most of the morphometric parameters. In this case, the maps and vertical sections—if digitized in convenient format—can refine the results. If the records are incomplete (e.g., the passage width and height data are missing), the maps and sections can be used to complete them. This is more easily achieved in a standard GIS program (e.g., QGIS).

The program and the file-folder structure, will be a part of the information system. If more programs and people are involved in the data processing, a unified nomenclature of files and thoughtful folder structure can help to avoid file access failures. The reliability of the resulting data depends mainly on the quality of the original survey, but due to the manual acquisition of data and sketching of passage morphology, it is always biased compared to the new methods [13]. Furthermore, the digitization of archive data is also prone to transcription errors.

3.2. Data types of new surveys

New surveys are still carried out mostly with the station-target approach, but with modern (fast and accurate) instrumentation. The accuracy and resolution of the collected data is usually much larger[1] compared to the traditional methods,[2] but still requires human expertise both in the data collecting and the processing phase. The most widespread surveying tool today is the DistoX, which is based on the combination of a laser measuring tool and a handheld computer [15]. The two devices are connected via Bluetooth, and the mapping program on the mobile device (PDA, tablet, or a smartphone) handles the database of the measurements providing graphical user interface for the on-site map compilation too. The software—running on the handheld computer—automatically handles the loop closures if new survey tracks are measured, modifying the coordinates of the existing stations too. The method is based on the algebraic minimization of the root mean squares (rms) of the differences.

[1]Precision of the Leica DistoX is 2 mm within 10 m range, with an angular error of 0.5° RMS [14].

[2]The spatial accuracy of the traditional measuring method is 1% of the distance from the entrance point in good conditions, but it can be even 10% [6].

Although, this mapping system is a GIS application in itself, it is not designed for the post-survey processes (e.g., map making or morphometry analysis). For these purposes, several external component programs are used that can import the surveying program's output file types. The output formats are the common vector graphics (e.g., dxf—a simple text type file describing shapes and geometry in a well-documented syntax [16]), and the text-type database with rows and columns are compatible with the usual cave survey managing programs. In the case of a DistoX survey, the raw data structure is quite similar to the previously introduced traditional database (**Figure 3**), having the advantage of being in digital form natively.

In contemporary surveys, the ultimate aim is to increase the speed and accuracy of the measurements using digital data. With DistoX, the transcription errors can be bypassed by direct recording in the handheld device [15], but there are still biases: (1) shooting the laser beam to a few selected points in the spelunker's field of view and (2) the manual generalization of the cave morphology by drawing the map on-site. Although, the process can be enhanced—shooting more and more targets—the resolution of the survey will always be lower than the surveys' done with a TLS.

The use of static terrestrial LiDAR instruments—despite of their impractical nature in harsh environments—is on the rise [5, 18, 19]. These tools produce thousands of range and angular data in few minutes measured from the station's location. The target points—similarly to the DistoX—are measured with one laser beam, but in this case the instrument repeatedly shots the beam to new targets swaying almost the whole field of view during one session. The point cloud of a scanning session at one station consists of nonoverlapping points forming a grid when using 3D polar coordinates (yaw, pitch, and range) or a data table when using Cartesian (x, y, z) coordinates [20, 21]. The former one is considered as a raster type data and can be easily fitted with panoramic photos shot from the same position. To do this, the scanning instrument should be equipped with an optical camera too. However, it is more common to export the scanned data with x, y, z coordinates in binary.las files [21]. Although, other formats are also exist, most of the point cloud processing programs (e.g., MeshLab, ReCap, Microstation, and CloudCompare) are able to import and export las-files.

Figure 3. The screenshots of the Pocket Topo program, which handles the data received from the DistoX tool [17]. The table view on the right shows the records of each measurement, the middle image is a map view displaying the connected stations, and the right one demonstrates the sketching options and the survey shots initiated from the stations.

Concerning the coordinates, the point cloud data is in a local reference system relative to the station. Data from multiple stations can be combined if the scanned surfaces overlap with each other, and if artificial backsight targets (i.e., regular-shaped objects) are placed into the common field of view of the subsequent scans. This process is done either automatically or manually within a desktop application after downloading the data from the TLS instrument. Both processes are based on "best fitting" approximations defined mathematically in the algorithms of the processing tools. The error of the fitting depends on the method we choose in the fitting approximation—usually the least-squares method—and the range of error is usually documented in the programs' description, although it also depends on several other factors. According to Lichti and Gordon [22], there are five error types which are distinguishable in a TLS survey: (i) the placement of the survey stations and the backsight target object; (ii) instrument leveling and centring; (iii) backsight target centring; (iv) raw scanner observation noise; and (v) laser beamwidth.

The precision parameters for the instrument, and the method can be obtained from the documentation if we know the range of the shots (beamwith of the laser beam is calculated) and the magnitude of the de-noise algorithm (removing outliers from the point cloud) relative to the range. Yet, at least two of the above listed errors are not independent from the human factor during the survey: the placing and the leveling of the instrument. However, attempts to reduce the chance of human errors are already made in the TLS procedure too; some LiDAR tools do not require manual fitting of backsight target objects to position themselves at the subsequent stations, and the precise leveling of the instrument is also done with automatic sensors and motors. Sometimes, though, this is quite problematic because of the size of the TLS instrument and the positioning is still the decision of the surveyor (**Figure 4**).

If the fitted sequence of the survey sessions contain at least one (but rather two: entrance and exit) positions where the geographic coordinates are measured with GPS (or other geodetic instruments), the whole survey can be transferred from a local (x, y, z) to a geodetic coordinate reference system (X, Y, Z) and can be referenced to other data (i.e., maps) surveyed previously (**Figure 5**).

Leaving behind the station-target method, techniques of high-resolution mobile mapping—such as the Zebedee—are also emerging [4]. Two data types are generated from this approach: a point cloud and a trajectory. The point cloud data is a huge list of x, y, z coordinates enriched with a set of attributes (the intensity of the reflecting beam or the precision of the calculated coordinate) associated to each of the points. The trajectory data is a much smaller set of coordinates in a strict sequence, defining the movement of the surveyor in the Euclidian space. This data is quite similar to the polygon network of the archive surveys—with the distinction that the vectors of the trajectory do not necessarily join in one single node at the branching points of the passages. The instrument is a lightweight handheld LiDAR station combined with an inertial measurement unit (IMU), which provides measurements of angular velocities and linear accelerations. The IMU also contains a three-axis magnetometer. Based on the incoming data from the measuring instruments, a portable computer calculates the trajectory of the surveyor and the position of the point cloud relative to this trajectory. With this instrument, several thousands of point data are collected within seconds; and obviously, the method,

http://dx.doi.org/10.5772/intechopen.68833

Figure 4. A TLS station positioned as close, as possible to the cave wall to "see" into a vertical chimney. The white spheres are the artificial backsight targets kept in position till the next session is measured from a subsequent station.

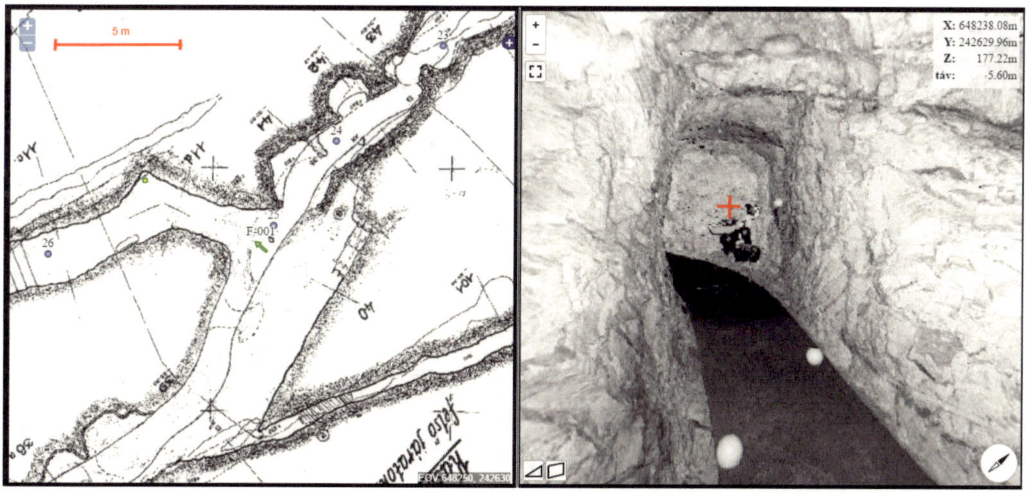

Figure 5. Archive map (left) and TLS survey image (right) in a common reference system. The map is oriented to the North. Green arrow indicates the position of the TLS station pointing toward the line of sight of the intensity image shown in the right panel. Geodetic coordinates of the station and the cursor position is displayed. The figure is created with a web viewer, which was designed to explore the TLS data beforehand thorough data processing [20].

which estimates the trajectory may produce errors. To correct these errors, the comparison of the overlapping areas helps—like the loop closure method in traditional cave surveying—to minimize the differences. The software uses best fitting algorithms to automatically localize similar patches of scanned cave parts [23]. The point cloud data—similarly to the data types of a TLS—is a .las-file or a zip-compressed .laz-file.

3.3. Combined data types of archive and new surveys

When new survey is done with modern instrumentation, usually the subject is a cave, where spelunkers worked previously and produced several kinds of archive data. The newly measured and the archive data both provide valuable information for scientist, thus, they should be integrated with each other. The two datasets can be paired along well defined spatial constraints—like identifiable morphology or artifacts (**Table 1**). For example, if some points of the archive survey are marked permanently in the cave, the installed artifacts can be identified on the LiDAR point cloud as regular-shaped objects. If the markings are too small, more apparent objects can be mounted temporarily on the cave wall, where old markings are found (e.g., uniform-sized disks).

In some cases, the structure of the archive dataset (i.e., column sequence in the data table) may have similar characteristics to the new one, but the reference systems of them are different. To avoid errors in later phases, the two sets should be checked at overlapping parts before unifying the two databases.

Archive data is not necessarily old data. Point clouds of several TLS survey sometimes are given to scientists to process the data and extract new information from it, but the surveys may come from different groups, who worked with different instruments. It is also possible that the point cloud data (las-files) are not accessible, only the 3D model of the cave—derived from the point cloud. Such models can be created in several ways—basically using stochastic methods—and the

Archive data	New data	Action
Map	Survey database	Identify station locations on the archive map
Survey database	Survey database	Match the data structure; harmonize the coordinate system (check the validity on overlapping areas)
Map	Point cloud	Identify characteristic points on the archive map (based on morphology, or permanent markings on the cave wall)
Survey database	Point cloud	Identify the possible locations of the archive stations—based on notes, and/or permanent markings
3D model	Point cloud	Compare the locations of the base stations; check the point cloud processing methods if only the mesh (3D model) is available
Documentation	Survey database	Depending on the type of the document: matching descriptions with the new survey, photo localization, section orientation

Table 1. Typical setting of various types of archive and new data types, and the necessary actions to put them into one data system.

generalization of the surface (i.e., the level of details) mainly depends on the used method. If the method is unknown for some reason, the relation of the 3D model and the original point cloud data has an uncertainty, and in practice, the correct position of the model can be achieved only if overlapping cave parts are present in both the model and in the newly measured point cloud.

Various documents may exist if a cave is well known for a long time, and the overarching aim of a GIS is to integrate these data into a *common spatial context*. Depending on the type of the archive data, the integration takes different amount of time. The process involves the same methodology, what is described in the case of the archive data (i.e., scanning and structuring), but it can reach better results due to the presence of the new survey. The most challenging task, however, is the positioning of archive photos, which requires deep knowledge of the subject cave. The archive photos can usually be located simply to one spatial position, but in some cases, both the subject's and the photographer's position can be reconstructed. This information is stored in a separate data table, or in the header of the image file, which can be extracted with a photo-editing program. Using this information, the photo can be draped on the cave surface model.

Not only archive photos are the subjects of the systematic process of cave-related tasks. Close-ranged photogrammetry is an emerging method of cave modeling besides TLS or combined with it [19]. While TLS is preferable if the morphology of a cave is the subject of the survey, while in the case of an archeological site, the texture of the cave wall is the prior aim of documentation. Even 3D models can be created simply from the images, if abundant overlapping photos are taken with the same interior parameters (focal length, distortion parameters). This is achieved with programs (e.g., Photoscan), which can reconstruct the relative spatial positions of the photographer via the comparison of the texture of the images.

4. The structure of a survey database

The GIS works effectively if the links between the different subsystems are well defined, and the users understand these definitions enough to maintain the links. The regulations should be built in the programs as deeply as possible (error handling) and be documented in manuals. The subsystems can be connected into a data system if:

- The data transfer is well established (documented).
- The quantitative data sets are in the same unit system.
- The qualitative data sets are defined on the same contextual basis.

The data transfers occur in those cases when the user attempts to work with data, which was created or stored in different program from which she/he uses momentarily. The most basic act of data transfer is opening a file with a program. If the file has improper syntax, this simple act may result in only error messages. Obviously, one can prepare for this by saving the survey data in proper file format, but what is "proper" depends on the software used as the component of the system. Ideally, the whole sequence of the processes is planned prior the survey, but it should be documented in written format at least during work; otherwise, the process is not reproducible. These documentations contain the list of the system components (*component programs (CP)*), the connections between the components (*input-output (IO)—formats*) and the location and the naming system of the files (**Table 2**).

In some cases—at least partially—the component program itself creates the documentation, which can be in various forms (e.g., text, rtf, and xml). In a component program equipped with GIS functionality, the user connects the different data types in a workspace, and the connections are saved in workspace (ws) file. In a general purpose GIS workspace such as the QGIS, an xml-type file is created containing the <datasource> tag (**Figure 6**). If the data-source file is renamed or moved, the data will not be loaded into the workspace, unless the user redefines the connection. This ws-file also contains the coordinate system, the querying, and the visualization parameters of the data source. Alternatively, the user can create log files, where one can track the changes made in the different system components—providing more control over the whole process—however, this requires great self-discipline, and such logs are very rarely published.

From the technical aspects, a well-written documentation is usually more valuable than the published result of the survey (i.e., a map or a 3D model of the cave) because it contains the stepwise methods, providing the benefit of reproducibility of the results. But writing a thorough description of the methodology not only benefits third party readers, but the scientist too who actually work on the project, because the documentation itself may also form the basis of scientific publications. In the process of reconstructing the results of an archive survey, the logs are also quite valuable.

4.1. Database management

Due to the modularity of the GIS, the survey database is neither a homogenous table (i.e., an Excel worksheet), nor a uniform file, although the data modules may take such common forms.

	Scope of the documentation	Description
1	Database management	Definition of database structures (RDB andxml), queries, connections, table/file names, attribute types, etc. Handling the different conceptual categories in qualitative data types (i.e., different nomenclature in source data). Handling the various measuring systems, units and calibrations in quantitative data types
2	Processing sequences	Sequence of the actions of the work (from measuring the data to publishing the results), and naming the programs, which were involved
3	Automated processes	List of scripts and programs that perform automatic processes, indicating certain actions they involve. Logic and syntax of file and folder naming for input and output files
4	Quality control	Description of the possible errors of each action (both manual and automatic), and the definition of the acceptable error range. Description of the possible validation methods

Table 2. Main contents and scopes of a GIS documentation describing the data structure of the system.

```
<maplayer minimumScale="0" maximumScale="1e+08" simplifyDrawingHints="0"
minLabelScale="0" maxLabelScale="1e+08" simplifyDrawingTol="1" geometry="Point"
simplifyMaxScale="1" type="vector" hasScaleBasedVisibilityFlag="0" simplifyLocal="1"
scaleBasedLabelVisibilityFlag="0">
      <id>interpolated_points20160909</id>
      <datasource>./modell_0m_5x5km_15km_500k/interpolated_points.shp</datasource>
      <title></title>
      <abstract></abstract>
      <keywordList>
        <value></value>
      </keywordList>
      <layername>interpolated_points</layername>
...
</maplayer>
```

Figure 6. Part of the xml-type QGIS workspace file (.qms) showing the syntax of the data-source definition (the relative folder route from the workspace file and the name of the data-source file are in bold).

Original form of the data modules are differentiated into *raster*, *vector*, and *alphanumeric* attribute types. If the original data is assigned into a workspace file, where the different types of data are linked, a GIS database is created despite of the diversity of original forms.

Database management involves the structuring, the maintenance, and the querying of the data through one or more user interface (program component) such as a GIS application (**Figure 7**). In a GIS, it is very rarely a linear sequence of tasks, but rather an iterative process. The iteration involves mainly the positioning of the data: if new measurements are available, the calculated coordinates of the existing—processed—data may change due to the fitting methods. Also, the quantity of the processed archive data may increase. To avoid discrepancies in the database, the relations of the different data modules should contain links pointing to each other. These links allow one to manage the system without changing the database records manually. This is usually done automatically within the program component that manages the different data types (e.g., closing a loop in Therion will modify the passage geometry of the whole cave map [3]).

The data physically is stored on a hard drive, and it is evident that the data access must be ensured throughout the managing procedures. To ensure this, the folder structure is recorded in the workspace file, but it is also important to notice that the component programs—installed on the computer—have their own logic in storing the data. For example, the default file saving path of a GIS program can be the same where the file settings were installed, and when the program is uninstalled or reinstalled with a new version, this folder can be overwritten or removed. These folders are the *system folders*, which belong to the component programs together with those folders where the executables are installed. To avoid loss of data, one should not store or save acquired data or workspace files in system folders.

4.2. Processing sequences

The well-documented sequence of actions in the work (from measuring the data to publishing the results), naming the programs which were involved, is like a cookbook: one can achieve the results without it with experience, but for those, who are not familiar with the whole procedure,

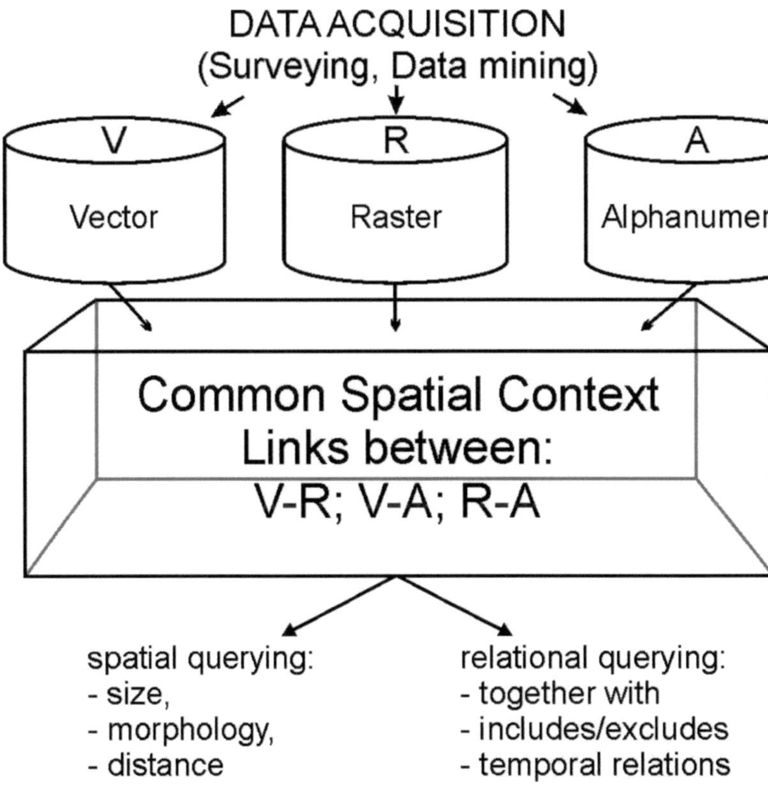

Figure 7. Flowchart of the database management in a GIS. The structuring is done when the acquired data is organized into different types of databases; the management includes the definition of the common spatial context, and the links between the databases. The querying tackles spatial, logical, topological, and temporal relations.

the stepwise aid is necessary. The processing sequences are the main components of a technical documentation. Basically, two scenarios can be distinguished:

- The data acquisition is already done and the GIS is aimed to incorporate the available information into a coherent system.
- The data is yet to be acquired, and the project can identify core components of the GIS, which are planned.

In the first case, the project obviously involves archive data even if the data is not so old, and the main task is to revise the different sources. The processing sequence describes what and how we modify the data to incorporate it in a GIS program. However, the data quality—in this case—cannot be truly modified, by filtering out the unmarked records or outliers, it can be enhanced. Modifications are done in the qualitative and quantitative categorization of the data (measure units, coordinate system, and nomenclature). The definition of qualitative categories may also be different in the data (e.g., what is considered as the minimum size of a conduit?).

In the second case, the planning should include the preparation of templates (e.g., tables), and user guides. Moreover, the incoming data should be error-checked, prior to passing it to the subsequent phase of the work.

4.3. Automated processes

In most cases, the automatic processes help the users in time consuming, repeatedly occurring tasks, such as recalculating the spatial positions of the survey stations after loop closure [3]. Not so long ago, the cave maps have had to be updated manually after a new survey track was added to the existing system, even if the loop closure was calculated by the software. Fortunately, this task is also automatized in some of the cave surveying programs [3]. The automatic methods in cave data processing are developed in three main areas:

(1) location determination

(2) modeling

(3) data management

Determining the spatial position is one of the most challenging tasks even in contemporary surveys. The polygon network ideally has to be connected to two distinct surface points (entrances or wells) to fit the lower order stations in between them, but the two way measuring can also improve the quality of the data. The fitting (calculation of the station's position) is done by linear algebraic equations and automatized in the surveying program. The logic is the same if we work with LiDAR.

Modeling of the cave is done for several reasons, but the aim is usually to create an irregular shape in the virtual space. Subsequently, one can calculate the parameters of it, or simply use it as a visualization of the hardly accessible location. The models thus, are *parametric* or *realistic* ones. Parametric models can automatically be generated from the survey records extruding 2D geometrical shapes along the station-target vectors [10, 11]. The visual representation of such a model is schematic compared to a realistic one (**Figure 8**). To produce a parametric

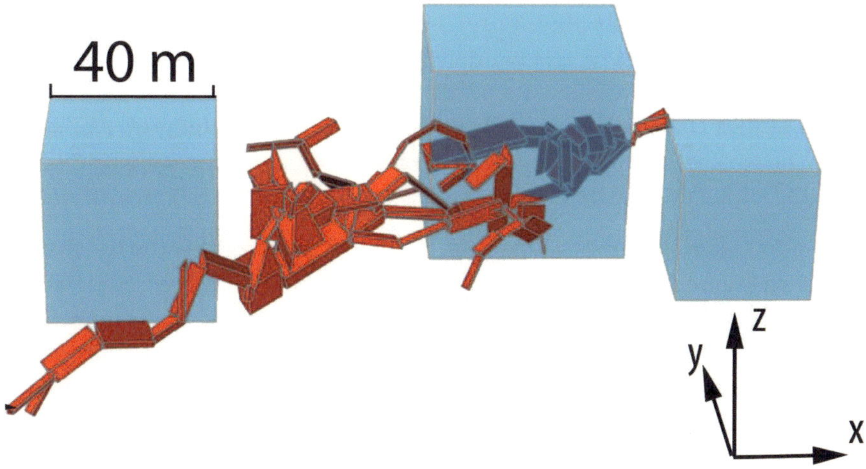

Figure 8. Volumetric model of the Szemlő-hegy Cave (Hungary) with regular-sized control objects (cubes). Such cubes can be used as references to calculate the relative proportion of the macro-porosity (passages) within a certain volume of the incorporating rock using statistical methods [11]. North is parallel with the y axis.

model, one must not even have to visualize the model to obtain the results, which are numbers indicating the volume, surface, and rate of void in the incorporating rock.

Creating realistic models—however—is a more common aim among cavers. Besides the table-, and the map view, the cave surveying programs provide 3D visual representation of survey results for a long time (helping the cavers to understand the passage structure more easily). In the popular surveying programs, the modeling is also based on the extrusion of a geometrical object along the station-target vector, but to enhance the model resolution the vertex number is increased in the surveys regarding the transversal sections. Instead of just 4 (LRUD method), 6 or 12 equally distributed radial vectors are measured around a station perpendicular to the station-target vector [7]. The sections are placed along the polyline network, and the more the vertices are measured, the better the realistic model will be. The edges of the adjoining shapes are also smoothed automatically using tangential curves and radial base functions.

When working with a LiDAR data, the automatic methods help fitting the point clouds of different stations to each other. The fitting result is described with statistical parameters and in some cases with a new attribute of each of the points showing the quality (*quality map*).

Automatic processes in data management are responsible mainly for the data loading and updating. This automatism occurs when the database is located on a server, while the GIS interface is on a client computer. If a working GIS is established and the database connections are defined, SQL scripts can update the client side regularly querying the database server. The data upload process is also automatized in this case: the data logged in a survey management program (or just in an Excel sheet) is written in a certain file format, which can be data-mined with scripts (primitive programs developed for repeatedly occurring tasks). The script code can work with any type of data (raster, vector, or alphanumeric). It extracts the data from the structured file and uploads it to the server-side database. The data mining scripts only work well if the files are located on the predefined path/folder; otherwise, the data is not loaded in the main database.

4.4. Quality control

One of the most crucial issues of archive data processing is the estimation of errors, which are present in the sources. Errors mostly affect the spatial positioning of the base data, thus, it is important to find ways to compare the existing data to something we can surely decide whether it can be trusted. It is also important to know how the errors were originally put into the data. The QC of the archive data is usually based on new (control) measurements.

For example, if loop correction was done with survey management software, the geographic position of the passages might have changed drastically. On the printed editions of the map, this was not always fully tracked. The farther in the past we go back, the bigger is the chance we find cave maps with uncorrected parts edited manually after new survey sequences. In fact, most of the result maps of archive surveys inevitably bear such kind of inhomogeneous errors distributed over the whole area.

This produces many possibilities for subsequent misinterpretation and first of all, we have to obtain a consistent database of the archive survey tracks to calculate loop closures. Though, this task is only a matter of digitization of the field notes in most cases, it is quite problematic if the survey database (records of the measurements) is not available. In the latter case, the polygonal network of the survey has to be reconstructed from the maps, and there is no way to tell what the error of the survey is, until conducting a new one. However, it is observed that the estimated error is not less than 1%, but sometimes reaches 5–10% of the distance from the base station [6].

Correcting the geometry of an archive cave map is often a first step toward building the GIS. Regarding the inhomogeneity of the spatial errors over the mapped area, the correcting method must also apply to spatially varying functions to modify the misplaced parts of the map. The simplest—and most adequate—method for this is based on the irregular network of triangles among ground control points (GCPs). The GCPs are usually the stations of the archive survey and can—ideally—be identified on both the scanned archive map and the line plot map, which is created from the survey database. The two maps make two differing geometric manifestations of each triangle in the network, although the corner points (stations) are literally the same. The comparison of the triangle pairs—using the Euclidian coordinates as variables—results a first order 2D function of transformation, which can be applied on all points within the area of the triangle (including the nodes). Usually the line plot map is accepted as the base and the scanned paper map is the one to be modified triangle-by-triangle. This method is also referred as the "rubber sheet" method. The absolute difference between the two positions of the triangle's points can be expressed as an attribute in the database making it possible to compile 2D "error maps" as quality indicator of the archive map.

After correcting it, even an archive map can be used to extract enough spatial parameters for 3D cave modeling. The volumetric parameters of the cave model can be calculated from the transversal and longitudinal sections. The error is estimated from the comparison of the archive section and a newly measured one. For this comparison, basically the geometrical parameters (perimeter, area) are used. The experienced rate of difference depends on the resolution of the source material (the lager, the better) and the shape of the passage profile. This implies a very important thing in cave data processing, which is the uniqueness of each cave.

In the case of new surveys, the QC is usually maintained largely by the processing programs by logging the accuracy of TLS measurements and the surface fitting parameters. However, the user must consciously handle these logs (often presented as a simple message after finishing a work phase) to track and report the confidence of the created model. In this case, the user created logs and documentation of the data processing can be the proper form of having a control on the quality.

5. Uniqueness of cave investigation

Caves are unique systems which evolved in unique geological, morphological, and climatic circumstances. Although following the processing sequences and using automatism may help one to process and analyze the data quickly, people have to keep it in their minds that the certain cave may be different from the previously processed one. The uniqueness on one hand comes from the differing aims, but it also originates from the specialities of the surveillance techniques. Concerning the surveying part of the investigation, the position of the cave relative to the groundwater level is the most influencing factor followed by the passage geometry. The surveying can be extremely difficult in subaquatic caves, where the water is muddy and easy in dry and comfortably wide passages with box-shaped transversal sections.

During the last decade, the TLS surveying technology has evolved to a level of flexibility that makes it adaptable to suit different geometric conditions in caves [24]. There also exist several well-documented projects, which provide the necessary stepwise help in combining photo documentation with TLS survey data. The uniqueness of the geological settings and the hydrogeological history of the area however, is still an issue in interpreting the data. Thus, purely mathematical approaches extrapolating throughout whole regions must be handled with caution because they may lead to false results. It was shown by several authors that on regional scale, the regularities of the cave distribution in the karst depends on variables like the thickness and dipping of beds [25], presence/abundance of tectonic fractures [26], the rate and direction of vertical movements [27], and the hydrological settings (hypogenic/epigenic conditions [28]).

6. Related sciences (convergence in scientific approaches)

GIScience in the context of cave investigations represents a convergence of surveying (geodesy), modeling (mathematics), and application of the results (different disciplines of sciences). The disciplines include mainly natural sciences (geology, hydrogeology, karstology, climatology, morphology, etc.), but archeology may also be involved if the cave contains cultural heritage. This highly varying scope of possible surveyors of caves brings up concerns about the reusability of the surveyed data. A spelunker should bear it in mind that the collected data will affect the cave in longer terms either positively or negatively. If the collected data is disclosed or unorganized, scientists of different disciplines may have to survey the cave over and over impacting the environment with each attempt. The importance of reworking the archives comes into front light especially in those cases when the cave environment has changed drastically (due to opening parts of it to public), but also in those cases when the environmental impact of a new survey is high.

In a case study of the Buda Thermal Karst System (Hungary), Albert et al. [11] demonstrated how to use a GIS to obtain new scientific results from archive data. This project aimed to estimate the macro- and meso-scale conduit porosity within the limestone and marl sequence incorporating the cave. The archive documentation included survey records, maps, transversal, and longitudinal sections. A method was worked out to create 3D passage models from the survey database using Visual Basic scripts and a GIS capable mapping application (AutoCAD), and subsequently extract volumetric data from the models. The database has to be prepared prior to the modeling, and the scripts provided the automation for the process of making 3D shapes from database records. In this case, the data information system included the original survey records but without properly measured LRUD (left-right, up-down) data. The missing information was collected mainly from analog maps. The study processed three caves of the Karst System and managed to estimate the unmapped size of the caves, concluding that one of the studied caves can be the largest of the country with 2/3 of its passages yet to be explored. A few years after the modeling was done when the prediction was confirmed [29]. The study was based on archive data without disturbing the protected cave environments with a new survey.

Caves attract not only scholars and explorers, but tourist as well. The caves are important sites for the public, and the stakeholders—when deciding about cave management—should rely on an information system that incorporates multidisciplinary observations [30]. In all cases, the aim is to minimize the damage in the caves and maximize the benefits of a survey. This can be achieved with a GIS afterwards, but a thorough planning of the survey is also important [31, 32].

7. The future of survey data processing

Technologies and methods that may be an integrated part of cave surveying in the future, have root in the present. The surveying techniques are changing fast, but the caves are still places where both the surveyor and the surveying instrument are challenged. The technological and ergonomic characteristics, the price and the handling of a new tool all should be optimal to reach a breakthrough and become widespread in cave surveying. As the DixtoX became popular a decade ago, and the TLS during the last decade, the emerging technologies like the LiDAR-based mobile mapping system [4], combined with close-range photogrammetry may take the place of the "most popular cave surveying" in the future.

Although, the mobile mapping systems are expensive and still unavailable for most cavers, technical requirements of close-range photogrammetry become affordable for wide public in the last decade. Even a mobile phone can be used to create a photorealistic 3D model of an irregularly shaped object, like a rock surface [33]. However, whole caves are not always suitable for photo documentation because of the casted shades of artificial lighting, contrasts and greatly varying distances. Archeological sites though are documented in several cases using the combined technology of TLS, a photogrammetry [19, 34]. From the aspects of Earth sciences, geophysical methods are also at hand to map the cavities and the lithology in the rock body that encloses a cave [35].

The seeds of the autonomous or semi-autonomous surveillance systems (robots and remotely controlled probes), which can combine the laser scanning technology along with other sensor types (magnetic, infra, sonar, gravimetric, etc.), are also present [36]. The tendency is toward the higher precision and the larger data size, and obviously the data management and the processing methods will also have to change to keep up with the higher demands. The increasing amount of data will demand for lager storage sizes if continuous recording is expected for hours, and the multiplication of sensors will demand for more power and space. Moreover, the system has to be ruggedized making it even larger. Contemporary fully autonomous systems—capable to navigate without GNSS (Global Navigation Satellite System)—are the size of a larger suite case and works only for a few hours. It is still a long time before these surveillance robots will autonomously do the caving instead of spelunkers. A rather probable option is the combination of semi-autonomous systems with the recent technologies. Using TLS for broad passages and drone swarms equipped with active sensors for high and tight passages will require human assistance, in positioning beacons for the swarm. Still these systems are in not even in planning phase at the moment.

8. Summary

This chapter gives an overview of the aims, benefits, and the possible issues of creating a GIS from cave survey data focusing on the data types on both the input (surveying) and output (modeling) side. The most widespread data types are listed and explained along with the functionalities of the system components. The here presented approach highlights that in cave investigations one does not use simply one program to process the data, but many of them (worksheet editors, map editors, and modeling tools). Although, in strict sense, not all of the used component programs have GIS capabilities, using them in a common project connects them into an information system which has to fulfill three functionalities: processing, storing, and representation. This chapter explains how these functionalities are handled in the case of new and archive data processing.

It is very crucial, and the chapter emphasize it in several ways that the archive data is precious despite of its poorer quality in spatial resolution compared to the data of recent surveys. The huge amount of archive data is lost if it is not processed and incorporated into a common information framework of the GIS. One should not forget the environmental impact of a scientific study when deciding about a new survey instead of data-mining the archives. In some cases, the type of cave management may also has changed since the time of an archive survey (e.g., the cave has been opened to the public), and some parts of the cave cannot be surveyed any more in its original, natural form. Without building such an information system from the archives, modeling and the related studies must rely only on the contemporary data.

The chapter can help cave investigations in two ways: for those who are already familiar with surveying it draws attention to the importance of procedures like data management, quality control, or automation; and for those who work with the data as beginner users, the paper can shed lights to the various tasks related to cave surveys.

Author details

Gáspár Albert

Address all correspondence to: albert@ludens.elte.hu

Department of Cartography and Geoinformatics, Eötvös Loránd University, Budapest, Hungary

References

[1] Cave TP. Information System. In: Proceedings of the XXIII International FIG Congress; Münich, Germany: International Federation of Surveyors; 2006

[2] Addison A. ArcGIS Geodatabase Data Model for Cave Science. Maryville, USA: Northwest Missouri State University; 2006

[3] Budaj M, Mudrák S, editors. Therion: Digital cave maps. 4th European Speleological Congress; 2008

[4] Zlot R, Bosse M. Three-dimensional mobile mapping of caves. Journal of Cave Karst Studies. 2014;**76**(3):191

[5] Gallay M, Kanuk J, Hochmuth Z, Meneely JD, Hofierka J, Sedlák V. Large-scale and high-resolution 3-D cave mapping by terrestrial laser scanning: A case study of the Domica Cave, Slovakia. International Journal of Speleology. 2015;**44**(3):277. DOI: 10.4311/2012EX0287

[6] Jeannin P-Y, Groves C, Häuselaman P. Speleological investigations. In: Goldscheider N, Drew D, editors. Methods in Karst Hydrogeology. London, New York: Taylor & Francis; 2007. pp. 25-44

[7] Pachos A. An Empirical Study of Cave Passage Dimensions Using Augmented Radial and Longitudinal Survey Data. USA: University of Akron; 2008

[8] Fish L. Computer modeling of cave passages. Compass & Tape. 2001;**49**:19-24

[9] Budaj M, Mudrák S. The Therion Book 1999—2016. Banská Bystrica, Slovakia: Speleoklub Banská Bystrica; 2016. 107 p

[10] Pardo-Iguzquiza E, Duran-Valsero JJ, Rodriguez-Galiano V. Morphometric analysis of three-dimensional networks of karst conduits. Geomorphology. 2011;**132**(1-2):17-28. DOI: 10.1016/j.geomorph.2011.04.030

[11] Albert G, Virág M, Erőss A. Karst porosity estimations from archive cave surveys—Studies in the Buda Thermal Karst System (Hungary). International Journal of Speleology. 2015;**44**(2):151-65. DOI: 10.5038/1827-806X.44.2.5

[12] Kárpát J. cartographer A Pál-völgyi-barlang. Budapest: Magyar Karszt és Barlangkutató Társulat; 1983

[13] Hunter D. A field trial of common hand-held cave survey instruments and their readers: Bullita Cave System. Caves Australia. 2010;**183**:10-12

[14] Heeb B. DistoX2 User Manual—Leica Disto X310 based DistoX. Firmware Version 2.4; 2015

[15] Heeb B, editor. Paperless caving—An electronic cave surveying system. IV European Speleological Congress; 2008. Vercors, France

[16] Burke P. DXF, Revision 12. [Internet]. 1998. Available from: http://paulbourke.net/dataformats/dxf/ [Accessed: 02 February 2017]

[17] Heeb B. Paperless Cave Surveying. [Internet]. 2016. Available from: http://paperless.bheeb.ch/ [Accessed: 02 February 2017]

[18] Gede M, Petters C, Nagy G, Nagy A, Mészáros J, Kovács B, et al. Laser Scanning Survey in the Pál-völgy Cave, Budapest. In: Buchroithner MF, editor. Proceedings of the 26th International Cartographic Conference. Dresden: International Cartographic Association; 2013. p. 905

[19] Lerma JL, Navarro S, Cabrelles M, Villaverde V. Terrestrial laser scanning and close range photogrammetry for 3D archaeological documentation: The Upper Palaeolithic Cave of Parpallo as a case study. Journal of Archaeological Science. 2010;**37**(3):499-507. DOI: 10.1016/j.jas.2009.10.011

[20] Gede M, Ungvári Z, Kiss K, Nagy G. Open-source web-based viewer application for TLS surveys in caves. In: Gartner G, Haosheng H, editors. Proceedings of the 1st ICA European Symposium on Cartography; Vienna: International Cartographic Association; 2015. pp. 321-328

[21] ASPRS. Las Specification. Version 14 – R13. Maryland, USA: The American Society for Photogrammetry & Remote Sensing; 2013. p. 28

[22] Lichti DD, Gordon SJ. Error propagation in directly georeferenced terrestrial laser scanner point clouds for cultural heritage recording. Proceedings of FIG Working Week; May 2004; Athens, Greece. 2004. pp. 22-27

[23] Bosse M, Zlot R, Flick P. Zebedee: Design of a spring mounted 3D range sensor with application to mobile mapping. IEEE Transactions on Robotics. 2012;**28**(5):1104-1119. DOI: 10.1109/TRO.2012.2200990

[24] Idrees MO, Pradhan B. A decade of modern cave surveying with terrestrial laser scanning: A review of sensors, method and application development. International Journal of Speleology. 2016;**45**(1):71. DOI: 10.5038/1827-806X.45.1.1923

[25] Filipponi M, Jeannin PY, Tacher L. Evidence of inception horizons in karst conduit networks. Geomorphology. 2009;**106**(1-2):86-99. DOI: 10.1016/j.geomorph.2008.09.010

[26] Palmer AN. Cave Geology. Dayton: Cave Books; 2007. 454 p

[27] Klimchouk AB, Ford DC. Types of karst and evolution of hydrogeologic settings. In: Klimchouk AB, Ford D, Palmer A, Dreybrodt W, editors. Speleogenesis: Evolution of Karst Aquifers. Huntsville: National Speleological Society; 2000. pp. 45-53

[28] Klimchouk AB. Hypogene Speleogenesis: Hydrogeological and Morphogenetic Perspective. 2nd ed. Carlsbad: National Cave and Karst Research Institute; 2011. 106 p

[29] Leél-Őssy Sz. Caves in the Buda Mountains. Landscapes and Landforms of Hungary. Springer; 2015. pp. 121-128

[30] Hale E. Protecting Oregon Caves. GIS Best Practices: GIS and Science. USA: ESRI; 2008. pp. 32-37

[31] Griffiths P, Ramsey C. Best management practices for palaeontological and archaeological cave resources. Australasian Cave and Karst Management Association Journal. 2005;**58**:27-31

[32] Gillieson DS. Management of caves. Karst Management. Springer; 2011. pp. 141-158

[33] Tavani S, Granado P, Corradetti A, Girundo M, Iannace A, Arbués P, et al. Building a virtual outcrop, extracting geological information from it, and sharing the results in Google Earth via OpenPlot and Photoscan: An example from the Khaviz Anticline (Iran). Computers & Geosciences. 2014;**63**:44-53. DOI: 10.1016/j.cageo.2013.10.013

[34] Nunez MA, Buill F, Edo M. 3D model of the Can Sadurni cave. Journal of Archaeological Science. 2013;**40**(12):4420-4428. DOI: 10.1016/j.jas.2013.07.006

[35] Karabulut S, Cengiz Cinku M, Tezel O, Dedecan H, Oygo A. Determination of Sinkholes with Different Geophysical Techniques; A Case Study in Yarimburgaz, , Küçükçekmece Lake NW Istanbul, Turkey. EGU General Assembly Conference Abstracts; 2016

[36] UNEXMIN. UNEXMIN project. [Internet]. 2017. Available from: http://www.unexmin.eu/the-project/project-overview/ [Accessed: 02 February 2017]

Section 2

Bats in Caves

Chapter 3

Bats and Caves: Activity and Ecology of Bats Wintering in Caves

Jan Zukal, Hana Berková, Hana Banďouchová,
Veronika Kováčová and Jiří Pikula

Additional information is available at the end of the chapter

http://dx.doi.org/10.5772/intechopen.69267

Abstract

Temperate regions of the world undergo a marked range of seasonal variation, most becoming extremely cold during the winter. Bats are the only group of vertebrates that have successfully exploited caves as permanent shelter. Although bats may use caves throughout all year, their most important role in ecology of temperate bats is as hibernacula. Here, we summarize various aspects of bat hibernation ecology, including variation in flight activity at the cave entrance; patterns of bat hibernation behaviour; site selection in hibernacula, including the importance of temperature during hibernation; and level of bat movement activity inside the cave. In addition, we review present knowledge on white-nose syndrome, one of the most important threats to cave-dwelling bats.

Keywords: Chiroptera, hibernation, flight activity, seasonal use, white-nose syndrome

1. Introduction

When one thinks about caves, the first image that comes to mind is that of a dark place full of stalactites and stalagmites, with lots of bats hanging on the walls. Bats are mysterious and scary creatures for most people but extremely interesting and enigmatic animals for zoologists. Not only their night activity, longevity, underground roosting, and active flight make them a fascinating species to study but the actual methods used to study them are also of interest [1]. Up to the 1990s, almost all bat research was closely associated with their roosts [2]; animals being captured at the roost entrances, measured, and marked. As bats have high roost fidelity, they can be here caught and recorded repeatedly [3]. Recent developments in ultrasound detectors and miniaturized telemetry, however, have significantly expanded the range of possible bat research topics to include subjects as time of foraging activity [1].

While microchiropteran bats are able to use a wide range of natural and mad-made structures as roosts, roost availability and presence of an abundant food supply are often the main limiting factors for bats, particularly in temperate zone. Roost availability can influence species distribution, foraging behaviour, social and mating behaviours, population size, diversity, and even bat morphology or physiology [4]. While providing many benefits (e.g. protection against bad weather and predators, effective thermoregulation, higher probability of mating and rearing young, lower foraging costs or information transfer), roosts also represent a major evolutionary pressure regarding the survival and reproductive success of each individual bats.

Bats spend a significant proportion of their life hidden in roosts, though their requirements may differ through the year or even at different times of the day. As such, the diversity of bat roosts is very high, ranging from short-term ephemeral to long-term permanent sites. Almost half of the approximately 1200 species of living bats, including all European bats, use permanent roost sites such as buildings, caves, mines, tunnels, tree hollows, or rock crevices [5]. Caves and similar underground spaces offer temperate bats long-term roost sites with specific microclimatic conditions that fulfil two crucial factors: a relatively stable above-freezing temperature (close to the mean annual surface temperature for the area) and high humidity [6].

In this review, we focus on the ecology of temperate zone bats roosting in caves of the Moravian Karst, Czech Republic (**Figure 1**), habitats that supply many of the bats' needs and that can be used year-round. In doing so we summarize the results of our research on various aspects of bat ecology over winter, including variation in flight activity at the cave entrance, factors affecting site selection within hibernacula, and level of bat movement activity inside the cave. In addition, we summarize present knowledge on white-nose syndrome (WNS). The

Figure 1. Map of Europe with the location of the Czech Republic indicated. The circle within the inset map indicates the location of the Moravian Karst.

study of these factors, along with a general understanding of bat hibernation, are essential prerequisites to understanding the impact of disturbance on hibernating bat populations and for providing focus to future conservation efforts [7].

2. Variability in cave use by bats (flight activity at the cave entrance)

The ecology and behaviour of temperate zone bats are fundamentally affected by seasonal changes in day length and other associated climatic variables [8], the effect of which become more pronounced at increasing latitudes. In order to remain nocturnal, therefore, bats must display behavioural flexibility in circadian and circannual activity patterns. We have been investigating nightly and seasonal changes in bat flight activity at the entrance of a natural karstic cave (Kateřinská cave, Czech Republic), an important hibernaculum monitored for hibernating bats since 1970 [9, 10]. Activity was recorded using a double infrared-light (IR) automatic logging system that allows discrimination between bats leaving the cave and those entering. Recently, automatic loggers capable of collecting large quantities of data over long periods are increasingly being used to monitor activity at European hibernacula, e.g. in the Netherlands, Denmark and Germany (e.g. [11–13]). The use of such IR automatic loggers has been shown to provide a reliable index of activity levels [14, 15] and, unlike netting, they have the advantage of not disturbing or interfering with the bats' normal activity. Their main drawback, however, is that they are unable to distinguish between individual bats or bat species [16, 14]. If the study is focused on the activity of the bat assemblage as whole, however, this is a minor problem. Connection of an IR logging system to a camera can help in later species identification, though the use of flashlight will affect natural bat behaviour. Note, however, that some authors (e.g. [15]) state that species identification using this method can be unreliable. Ultrasound bat detectors can also be connected to IR logging systems and these have been used to monitor activity of a single species (e.g. the lesser horseshoe bat *Rhinolophus hipposideros* [17], the greater horseshoe bat *Rhinolophus ferrumequinum* [18]) or overall activity of all species in a locality (e.g. [14, 15, 19]). Unfortunately, this method is not very reliable at distinguishing echolocation calls of individual *Myotis* species [20].

The level of bat activity (at the cave entrance) varies seasonally and five periods have been defined (**Figure 2**), all showing a non-random temporal distribution with flight activity concentrated around a specific time [21]. In each case, activity level is influenced by a range of climatic factors, the effect and contribution of which change nightly and over the year [22].

(1) *Hibernation period (mid-November–early March).* Bats show very low or almost no activity and departures from the cave are very rare. Interruption of lethargy in these bats is most often caused by (i) changes in ambient conditions outside cave, (ii) changes in the physiological state of the hibernating bat (e.g. dehydration), or (iii) direct disturbance [23, 24]. During hibernation, average temperature and daily temperature range (i.e. the difference between daily maximum and minimum temperatures) are key factors predicting the general level of flight activity [16, 18, 22]. As temperature increases, so the percentage of nights with bat activity also increases. Similarly, an increase in temperature fluctuation during the day will also

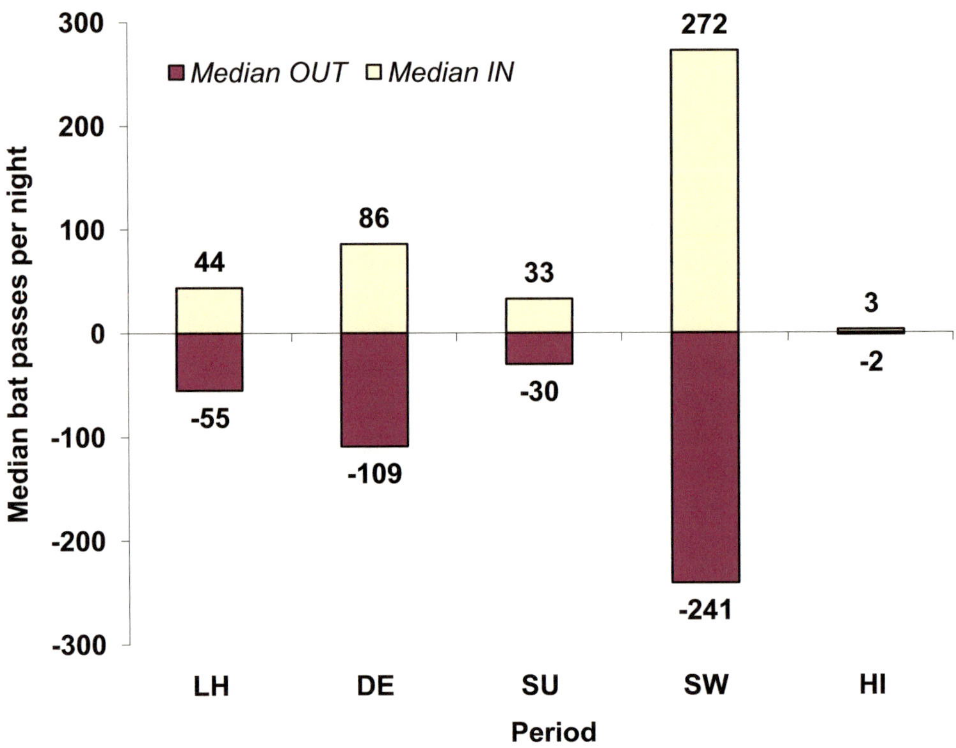

Figure 2. Out-flight (negative values) and in-flight (positive values) medians for each defined period as monitored by double IR-light barrier between March 2000 and November 2002. Explanation: HI, hibernation period (mid-November–beginning of March); LH, late hibernation (March–mid-April); DE, departure (and transition) period (mid-April until beginning of June); SU, summer period (mid-June–end of July); and SW, swarming period (late July–mid-November).

result in bat arousals and increased flight activity. Note, however, that bat activity at the cave entrance has been recorded at temperatures as low as −13.2°C (cf. [25]). Daily recordings were positive at maximum daily temperature exceeding 6.2°C, when some bat species are able to forage [16]. The activity within defined temperature groups [22] was significantly lower during deep hibernation period than during late hibernation (**Figure 3**). Opinions on the level of activity desynchronization at sunset and loss of nocturnality during hibernation differ and the results of research are inconsistent, some supporting desynchronization and others not (e.g. [18, 26–28]). Our own data [22] clearly indicate that activity at the cave entrance is synchronized with sunset, even in winter, and that a concentration of activity occurs between 3 and 3.5 h after sunset. No change in activity patterns has been recorded following the emergence of white-nose syndrome (WNS) in Europe, suggesting that the hibernation behaviour model described, including changes in activity, could represent a behavioural adaptation that has prevented fatal impact of the disease observed in North America [29].

(2) *Late hibernation period (March–mid-April),* with intensive departure activity during the first quarter of the night. Movement activity inside the cave is relatively high and the bats are

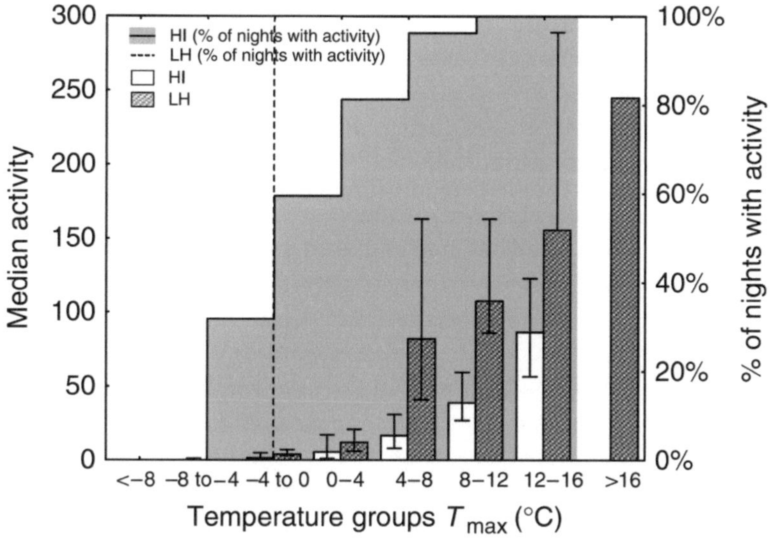

Figure 3. Activity levels (median values ± interquartile range) in individual temperature group during hibernation (15 November–4 March) and late hibernation (5 March–14 April). The percentage of nights on which activity occurred in individual temperature groups is indicated by the grey area for the hibernation period and the dashed line denotes late hibernation. *Source*: [22].

probably already preparing themselves for departure from the hibernaculum [30]. Flight activity is positively affected by average daily temperature, and negatively so by minimum temperature during the preceding day. Bats react very quickly to temperature changes from day to day, with activity decreasing or increasing if temperatures drop or rise by more than 2°C. Such rapid changes in activity level become feasible as the bats move towards the hibernaculum entrance, enabling them to register fluctuations in ambient temperature [19, 30] and, as a consequence, potential changes in insect abundance. Bats are capable of foraging at very low temperatures, e.g. Daubenton's bat *Myotis daubentonii* at temperatures as low as −3.3°C [31]. In some species, the activity increases during late hibernation period, presumably, as food availability is already higher and foraging effectively compensates for any energy loss [32].

(3) *Spring migration (mid-April–early June)*, a period of relatively high activity. At this time, the cave may serve as a transitional roost during the spring migration and, from around May, as a temporary roost for males as females already start to form summer colonies. Emergence activity in the first quarter of the night is high, and a small number of bats may re-enter the cave in the last part of the night. Average daily temperature and average daily atmospheric pressure at this time has a significant positive influence on overall flight activity. The degree of variability in activity explained by such climatic factors is the lowest during this period, however, suggesting that either temperature is no longer a limiting factor, or that endogenous rhythms have a strong influence on departure from the hibernaculum [11, 22]. However, the use of underground roosts including caves in the spring may be species specific; it may differ by region, and can also depend on roost structure [6, 18, 33].

(4) *Summer period (mid-June–end of July)*. During this period, the cave is used only sporadically (**Figure 2**), though the bats visiting the roost stay the whole night, i.e. they enter before midnight and leave after midnight. This type of activity suggests that, during this period, the cave may be being used as a night roost between peaks in foraging activity or as a transitional day roost [11, 18]. At this time, the cave entrance is visited almost exclusively by males [34, 35] as adult females occupy maternity roosts during lactation and return to these between foraging bouts, night roosts being used sporadically and for brief periods [36, 37].

Flight activity at the night roost entrance is influenced by fluctuation in ambient temperature, rather than any absolute temperature threshold, the higher the difference between maximum and minimum daily temperature, the higher the activity level. This corresponds with a model proposing that activity changes in temperate insectivorous bats reflect changes in insect activity [8], i.e. if day-insect abundance is high due to warmer nights, bat foraging activity may continue overnight with no visits registered at the cave entrance (low activity). On the other hand, when nights are cooler and the daily temperatures range is higher, bats will tend to spend more time in the night roost. Foraging activity is highest at dusk and just before dawn, after which the bats return to the day roost [36]. This model is also supported by the influence of rainfall, with flight activity at the cave entrance increasing as rainfall increases whether the nights are warm or cold.

(5) *Autumn migration or swarming period (late July–mid-November)*. This period is typified by very high general activity and an increasing number of bats entering the cave. With the break-up of the summer breeding colonies, activity at the cave entrance gradually increases as adult females and juveniles arrive [9, 38], often in small groups of 2–12. The majority of bats does not roost in the cave and probably arrive after the first foraging period; hence, peak activity tends to occur around midnight. Activity around the cave entrances in autumn probably enables juveniles to recognize potential hibernacula and to meet individuals of the opposite sex, which live separately during summer (e.g. [15]). Activity level is positively related to average daily temperature, atmospheric pressure and rainfall. Thus, when nights are warm and insect activity is high (high atmospheric pressure), the bats will quickly catch enough prey and will search for the cave entrances (swarming sites) in order to mate or obtain shelter it be raining [14, 22].

3. Caves as hibernacula

Hibernation, an optimal adaptation to a prolonged fall in temperature and reduction in prey availability, is a characteristic of the annual cycle of insectivorous temperate zone bats [16]. Selection of a suitable hibernation site is crucial for overwinter survival and, in temperate zone, caves and mines tend to be the most common hibernacula. Caves can be divided into three basic types based on microclimate and use by bats: (1) warm caves used during the summer, including maternity colonies, (2) relatively cold hibernacula with a stable microclimate and (3) caves used during the autumn swarming [6]. Of course, both warm caves and hibernacula can also be used during the spring and autumn migrations too. At higher latitudes, cave temperatures are too low and they tend to be used mainly during torpor and hibernation. Note, however, that while thousands of bats can hibernate at such sites, those sites with lower numbers may be very important locally and their overall contribution to bat population great [7].

More than 1200 caves are located in the Moravian and Javoříčský Karst regions of the Czech Republic, many of which host significant and regularly monitored bat hibernacula (**Figure 4**). Three of these cave systems (Javoříčské, Sloupsko-šošůvské and Býčí skála) represent the largest bat hibernacula in the Czech Republic [39], with 17 bat species registered during hibernation, including rare species such as *Rhinolophus ferrumequinum*, the northern bat *Eptesicus nilssonii*, and the pond bat *Myotis dasycneme*. A similarly rich bat fauna has only been found in caves in the Slovak Karst and the Muránská Planina [40], both of which are also located along the northern distribution border of some bat species (e.g. Geoffroy's bat *Myotis emarginatus* or the lesser mouse-eared bat *Myotis blythii*).

Both of these karst systems have a long history of bat research, beginning with speleological research of caves made by Dr. Friedrich Anton Kolenati in the second half of nineteenth century [41]. Modern bat research in the region was initiated by Prof. RNDr. Jiří Gaisler in the 1950s and it continues, including our long-term research of bat hibernation, to the present day. As a result, some of these hibernacula have been monitored for almost 50 years [42].

Figure 4. Main entrance of Sloupsko-šosůvské cave representing one of the largest bat hibernacula in the Czech Republic. Photo by Leos Stefka.

As one of the main requirements of our own research was to avoid any disturbance to hibernating bats, we used visual censuses only (including night censuses using Pathfinder 2000s night-vision scope) with no handling or marking [10, 29]. Thermal profiles were also undertaken to evaluate physiological condition. Fur surface body temperature, which is correlated with core body temperature, was measured using a Raynger MX2 non-contact IR thermometer (Raytek Corporation, USA). Two major model species were regularly monitored in the caves, the greater mouse-eared bat (*Myotis myotis*) and the lesser horseshoe bat (*R. hipposideros*) (**Figure 5**), these being typical members of the bat community hibernating in the Moravian Karst [9, 10, 43].

3.1. Model of bat hibernation in natural caves

In late summer and early autumn, bats undergo a preparation phase for hibernation during which they rapidly accumulate body fat deposits [44] needed for surviving the torpor period. The fat is accumulated by energy savings achieved through increasingly longer daily torpor bouts during the diurnal resting period. Hibernation is usually interrupted by periodic arousals [45, 46], usually related to drinking, feeding (in mild periods) or even mating [23, 35]. As part of the fat deposits must be metabolized for torpid individuals to become physiologically active during winter, such arousals are energetically costly [47, 48].

These arousals, and any subsequent activity, will be mirrored in ecological parameters such as community structure, bat population abundance, shelter selection or total movement activity. Monitoring of hibernating bats in the Moravian Karst has confirmed that the ratio of 'visible' bats changes through the winter, i.e. bats may move from inaccessible shelters to places where they can be monitored by investigators [9, 49]. The total number of hibernating bats grows continuously from October, with highest abundances occurring in February or

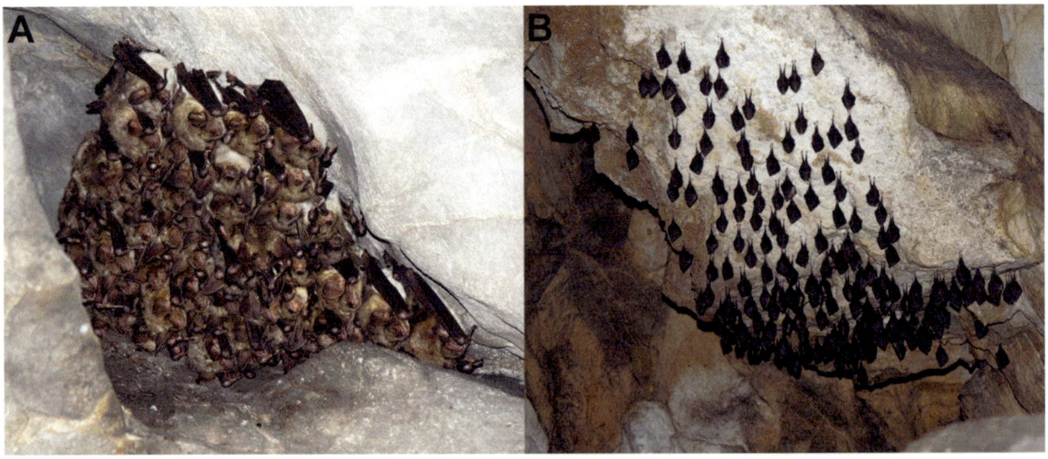

Figure 5. Hibernating clusters of two bat species regularly monitored in the Moravian Karst. (A) The greater mouse-eared bat (*Myotis myotis*) (body length of 6.5–8 cm) and (B) the lesser horseshoe bat (*Rhinolophus hipposideros*) (body length of 3.5–4.5 cm).

March, depending on community structure. Any increase in abundance will be influenced by immigration of newcomers during the pre-hibernation period only (mid-November–mid-December). Switching of hibernation sites during the deep hibernation period (i.e. leaving the hibernaculum) has only been registered exceptionally [21]. In April, there is a gradual but relatively rapid emergence from the hibernation sites (approximately 3 weeks), with bat abundance in cave decreasing to a minimum.

Movement activity of bats inside the hibernaculum, expressed as the percentage of new findings during a visit, is registered throughout the winter, with levels fluctuating in our species-specific models. Hibernation activity of *R. hipposideros*, for example, could be divided into three distinct periods reflecting early, deep, and late hibernation; while *M. myotis* movement activity remained relatively high throughout the season [30]. A continuous arrival of bats at the hibernaculum means that *R. hipposideros* abundance increased gradually over the 6–8 weeks leading to mid-December, and decreased again from mid-March as they gradually left (**Figure 6**). The deep hibernation period was characterized by low movement activity in the cave and minimal changes in abundance, as also confirmed by detection of ultrasound signals [17, 27]. Even in the middle of winter, when the conditions outside were suitable, some awakened *R. hipposideros* became aroused and left the cave, shortly to return again [17, 50].

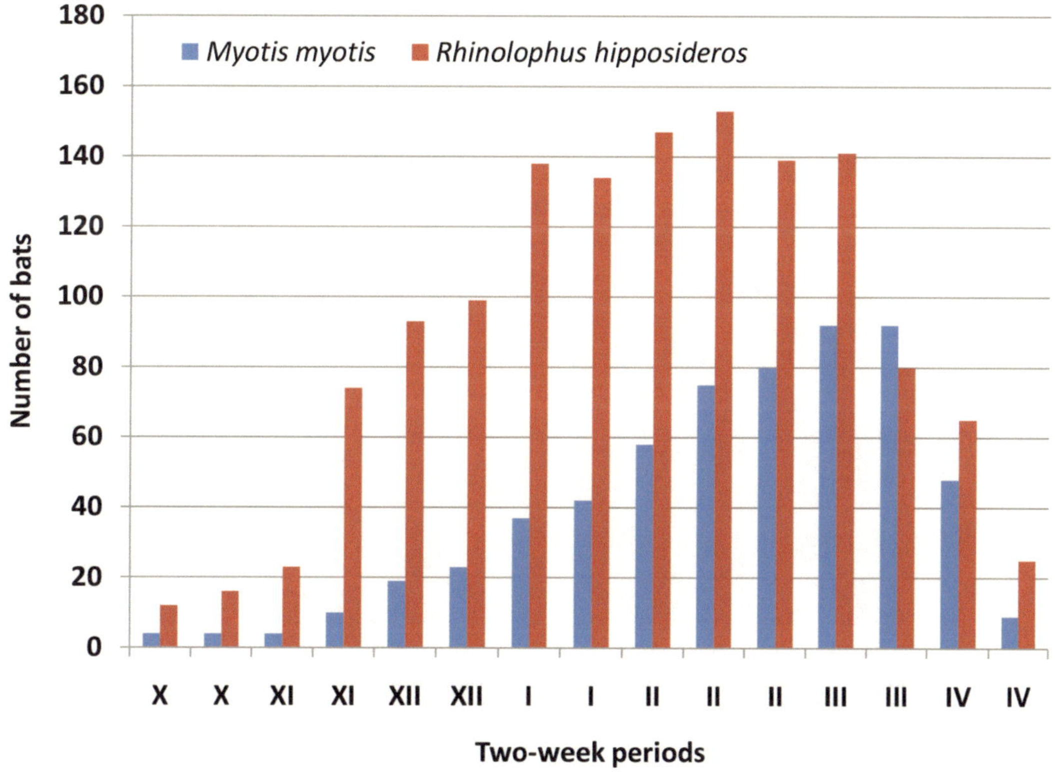

Figure 6. Changes in abundance of two model species during hibernation period 2011/2012. Data presented in 2-week periods.

Abundance of *M. myotis*, on the other hand, increased continuously throughout the winter, eventually dominating the bat community by the end of hibernation period. In comparison, this species leave the hibernacula rapidly, all bats having disappeared over a period of just 3 weeks [49, 51].

Our two model species accounted for more than 80% of all bat observations in the caves. Bat netting at the cave entrances during spring and autumn migrations, however, confirmed a much higher diversity than during hibernation, with other bat species showing a higher dominance. Small species of genus *Myotis*, such as *M. emarginatus*, *M. daubentonii*, Natterer's bat *Myotis nattereri*, and Bechstein's bat *Myotis bechsteinii*, are often underestimated during winter monitoring [9] as they tend to use more or less inaccessible roost sites (e.g. deep crevices) [34], depending on the local microclimate, species-specific requirements, season or weather. We found that around 20% of all bats hibernating in natural caves need to be monitored during winter as the cumulative number of bats entering the cave (calculated using a double IR-light logging system) was much higher (**Figure 7**).

3.2. Shelter selection during hibernation

As roost site characteristics can play an important role in bat thermoregulation, choice of site will undoubtedly influence bat fitness and survival. Ransome [52] classified caves used

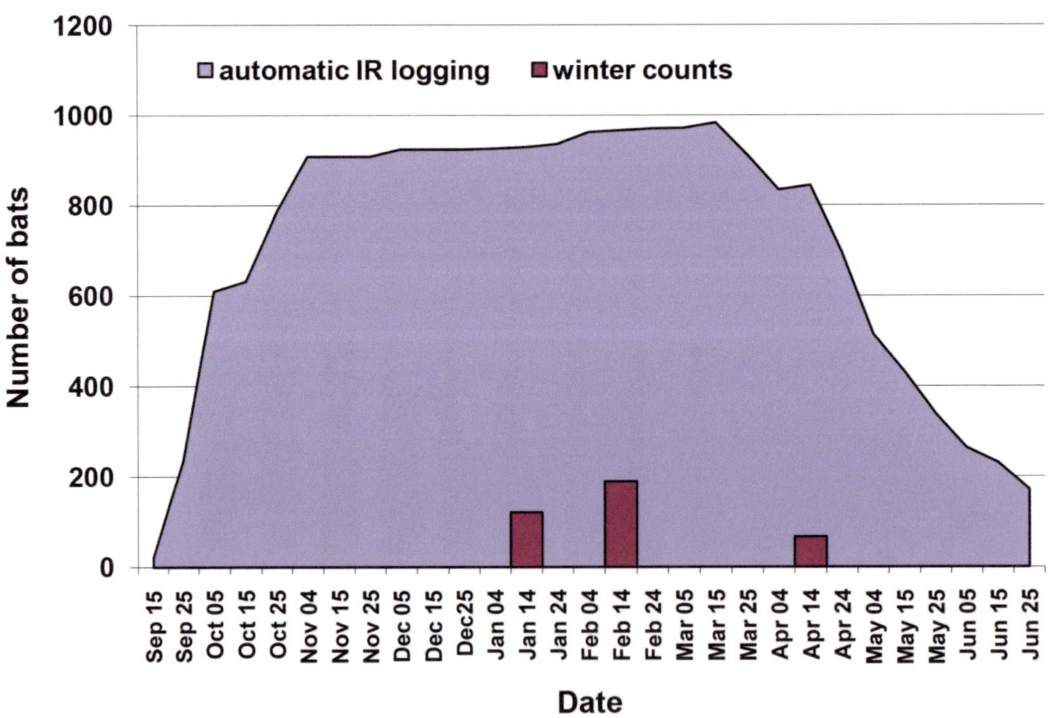

Figure 7. Cumulative number of bats entering the cave (winter season 2000/2001) recorded by the automatic IR-light logging system (area) and the numbers of bats hibernating inside the cave monitored during winter counts (bars).

as hibernation sites into three basic types depending on temperature fluctuation: (1) caves displaying a constant temperature regime, (2) caves with dynamic temperatures and (3) caves with fluctuating low temperatures. Note, however, that numerous factors affect the climate of individual caves; and that each cave will be unique in its geomorphologic and microclimatic parameters [6]. Caves with more or less constant temperatures over the year (averaging between 6 and 10°C) usually have just one entrance and temperature fluctuation tends to occur in the outer entrance parts only due to high air flow. Thermally dynamic caves are characterized by large passages with different temperatures. Such caves tend to have two or more entrances, their mutual positions influencing internal temperature conditions. As any two caves will differ significantly, therefore, it will be difficult to specify an average annual temperature. In general, average annual temperature will be in the range of 3–14°C. Hibernating bat communities sheltering in such caves tend to show the most stable abundances. The third cave type always tends to display fluctuating temperatures, despite usually having just one entrance. During winter, air temperature will decrease significantly due to cold air flowing in from the cave entrance [53, 54].

Survival of hibernating bats will be influenced not only by the selection of a suitable hibernaculum but also by the specific microhabitat conditions within. The correct choice will be crucial for the efficient use of stored energy and for the appropriate timing of flight activity. Indeed, studies have shown that bats are able to regulate length and depth of torpor by selecting favourable sites [46]. During our own monitoring of hibernating bats, we monitored a range of parameters including site type (exposed, semi-exposed and hidden), relative height above the floor and position in the cave [30]. During hibernation, *R. hipposideros*, a thermophilic species, were registered in practically all parts of the caves under study (Kateřinská and Sloupsko-šošůvské), with the exception of the entrances, which have more dynamic microclimates [30, 49]. The bats tended to prefer low shelters (under 3 m from the floor) and always hibernates hanging free in open unprotected sites, regardless of season or hibernaculum type. The sites selected by *R. hipposideros* had stable temperature and humidity conditions with minimal air flow and, as a result, the species showed very low movement activity levels during deep hibernation. The pre- and post-hibernation periods, on the other hand, are typified by high movement activity. Changes in shelter types during winter corresponded to phases in the annual cycle and the physiological status and behaviour of bats only and not to any changes in the environment, there being no temperature fluctuation in the deep cave sections used by *R. hipposideros* [16, 55]. Despite hibernating next to a footpath frequently used by tourists or speleotherapy patients, we failed to register any vulnerability of *R. hipposideros* to human activity.

Euryvalent *M. myotis*, on the other hand, were registered throughout the cave systems during hibernation, using all shelter types indiscriminately (exposed and hidden, ceiling and walls) and showing high seasonal dynamics. During deep hibernation (mid-December–early April), these bats are continuously moving into the outer parts of the cave where they select specific sites for the formation of clusters (**Figure 8**). Over 80% of all *M. myotis* hibernating in the Sloupsko-šošůvské caves, for example, were found in one specific area during late hibernation [49]. A shift towards the cave entrance has also been reported in other European hibernacula [56, 57]. Movement activity of *M. myotis* was relatively high in hibernacula throughout the hibernation period and could not be divided into specific periods [30]. In the absence of

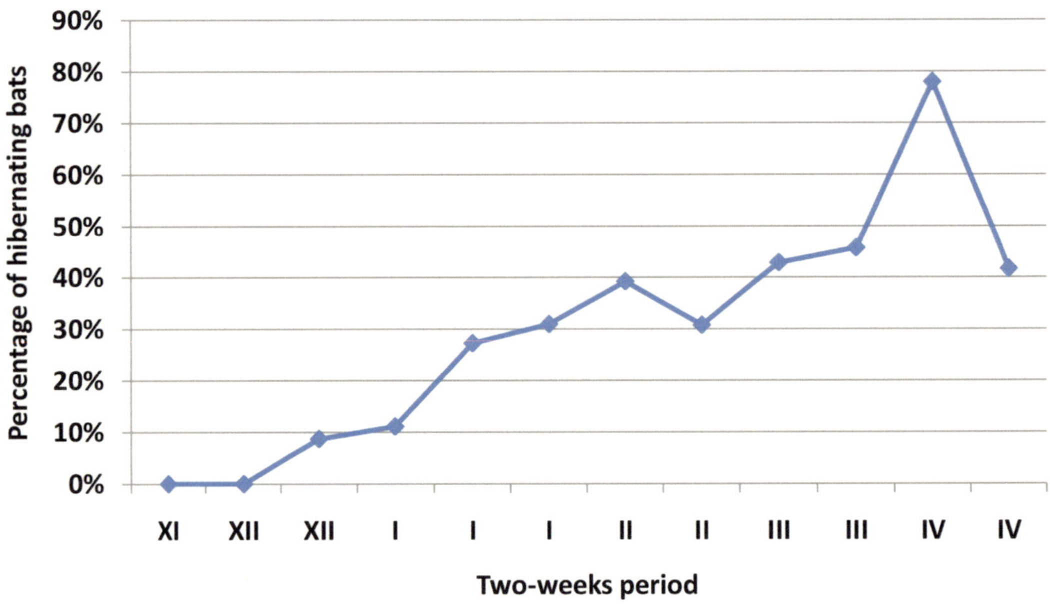

Figure 8. Percentage of the greater mouse-eared bats *Myotis myotis* hibernating in the Kateřinská cave entrance during winter 1992/1993. Data presented in 2-week periods.

food, *M. myotis* select sites with a constant temperature for deep hibernation in order to maximize energy savings. On the other hand, the bats most likely shift to sites with more dynamic temperature regimes during the late hibernation period as the changes in ambient temperature help bats synchronize arousals with actual weather conditions. In doing so, aroused bats are able to couple emergence activity with favourable climatic conditions for foraging.

It is apparent that neither all individuals nor all populations have the same model of hibernation [58]. Our studies suggest that *M. myotis*, at least at population level, may not follow the same hibernation model and that a range of hibernation strategies (i.e. level of movement activity, preference for different shelter types) may be used depending on the prevalent different microclimate profile (i.e. dynamic or stable). Populations in different hibernacula will exhibit responses tuned to that environment, while individuals of the same species may vary in the strategy used to survive hibernation. In this way, the bats optimize their depletion of energy reserves and improve their chances of surviving in the winter [59]. High fidelity to particular underground shelters also suggests that the adopted hibernation strategy may limit bats to repeated use of particular hibernacula [60].

3.3. Cave temperature and bat hibernation

The length of time that temperate bats can survive without feeding will be dictated by the temperature, at which they hibernate. In general, by hibernating in caves where temperatures are low but above freezing (i.e. between 2 and 5°C), the bat's metabolism rate is maintained at an efficient level. While the actual temperatures at which different bats hibernate is species specific [61, 62], the interspecific differences are very small due to the low metabolism and small body mass of temperate bats. Such species-specific differences vary seasonally, being somewhat smaller during deep hibernation and greater during the pre- and late hibernation periods. Bats also display intraspecific variations in preferred temperature, as individuals will select locations based on their energy reserves [63].

Bat arousal may occur as a result of temperature changes in hibernacula, following which the bats may move to a more suitable location [64]. In general, bats prefer to start hibernation at sites with higher temperatures as those with low temperatures may reach freezing point over the coldest months. An optimizing strategy of such type has been observed in *M. myotis* in natural karstic caves (**Figure 9**). As bats often return to the same sites year-after-year, this could suggest the use of prior experience, learning from others, and/or olfactory clues in microhabitat selection. Arousals are also temperature dependent, with the length and frequency of bat arousals increasing with temperature increases over 10°C [46].

During hibernation, bat body temperature falls to within 1–2°C of ambient temperature and metabolic processes slowdown, thereby reducing energy requirements. As a result, hibernation incurs physiological costs, including the build-up of metabolic wastes, dehydration, reduced motor function, altered immune response, and sleep deprivation [65]. Hibernation may also impose ecological costs such as decreased detection and response to predators [66] and an increased likelihood of freezing [67]. At the cellular level, cold stress changes cellular membrane lipid composition and suppresses the rate at which protein synthesis and cell proliferation takes place [68]. We examined

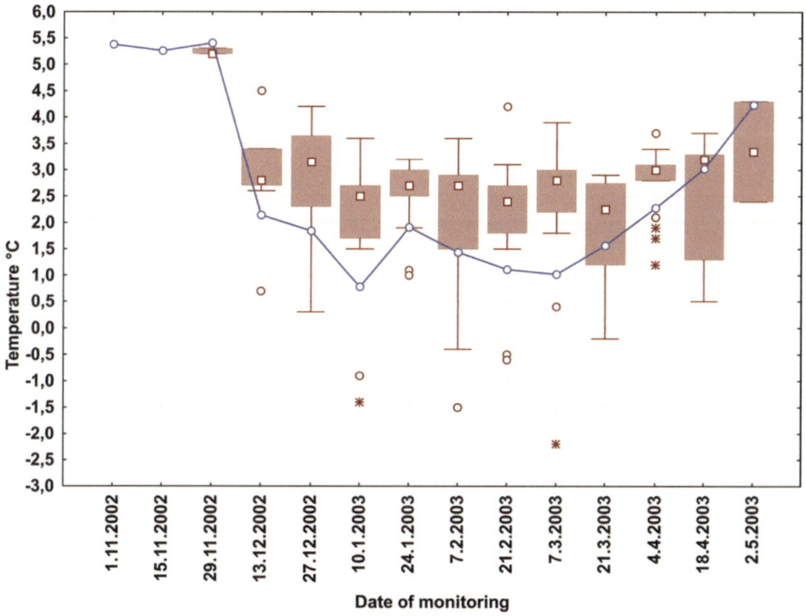

Figure 9. Changes in greater mouse-eared bat *Myotis myotis* body temperature and shelter temperature during the winter of 2002/2003. Explanations: box = interquartile range; middle point = median; whiskers = non-outlier range; circles = outliers; stars = extremes; continuous line = average shelter temperature.

the ability of primary skin fibroblast cells from the flying membrane of a hibernating *M. myotis* to proliferate under torpor and euthermia. After loosening the tissue mechanically (without proteases), the cells were identified as fibroblasts based on their spindle shape, positive staining for the vimentin mesenchymal marker, and the presence of typical stress-fibre organization in the actin cytoskeleton. Cell numbers for the assay started with 20,000 cells per well and these were incubated at 9 or 37° C for 6 days in a 5% CO_2 humidified environment for the experiment. Cells were detached from the cultivation wells and recalculated daily with 30 times repetition. While bat fibroblasts cultured at 37°C were elongated reached high numbers in 6 days, and attached successfully to the well substrate; those cultured at 9°C were spherical, reduced in number and took time to attach (**Figure 10**). Extrapolation from this cellular *in vitro* model suggests that bat fibroblasts have some proliferative capacity at the temperature conditions prevalent during torpor, though wound healing capacity would be much slower than in euthermic animals. Such a physiological response of bat cells may help explain the movements registered in *M. myotis* at low fur temperatures (T_{flow} < 5°C) [69], which would allow bats to save energy long-term and prolong torpor bouts. All T_{flow} events were recorded during late hibernation, when bats are faced with an acute shortage of energetic reserves and enormous metabolic requirements. In most cases, T_{flow} events were represented by slow displacements between clusters of bats, though departure or arrival to and from clusters was also recorded with no elevation in body temperature (**Figure 11**). Repeat appearances suggest that T_{flow} movements may represent a regular part of bat hibernation tactics.

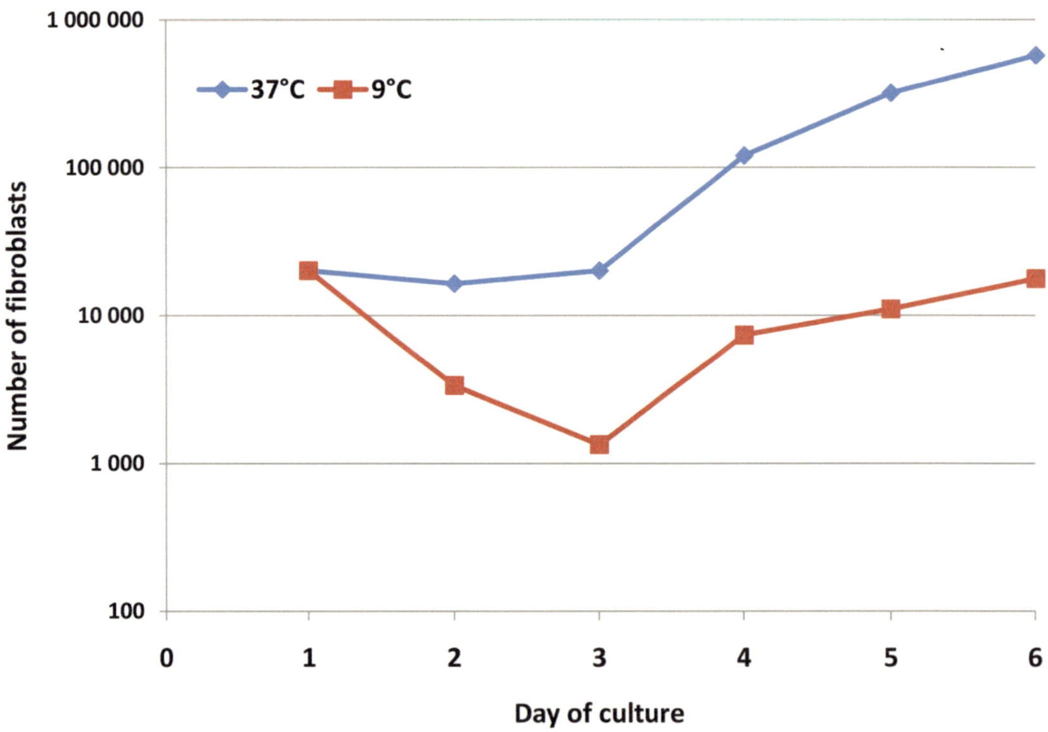

Figure 10. Numbers of primary cell (fibroblast) cultured at two different temperatures showing some proliferative capacity at the temperature conditions prevalent during torpor.

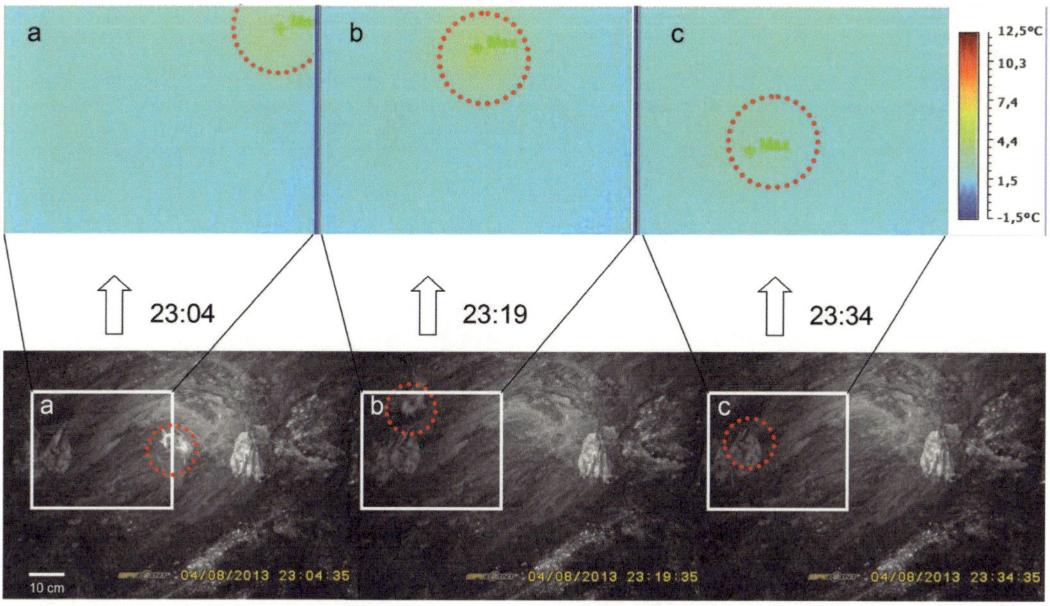

Figure 11. Examples of low body temperature movements, showing the bat moving between clusters from left to right. The upper thermal images correspond with lower images from photo-traps recorded simultaneously (a–c). The rectangles in the lower photo-trap images indicate the position of the thermal image, while the dotted circles indicate the moving bat. *Source*: [69].

4. White-nose syndrome: a threat to bat populations hibernating in caves

Bats are threatened by a range of both natural and anthropogenic stressors, including predation, lack of food, pathogenic agents, climate change, habitat loss, ecological disasters, illegal trade, chemical and light pollution, roosts and hibernacula disturbance, and wind turbine construction. Considering their economic importance to agriculture, the general decline in bat populations documented around the world is of some importance [70]. Recently, a novel threat to insectivorous bat species hibernating in caves and mines has been recognized in North America [71, 72]. White-nose syndrome (WNS), a fungal infection characterized by fungal growth on the bat's muzzle (**Figure 12A**), has caused a dramatic decline in American bat populations. *Pseudogymnoascus destructans*, the causative agent, is a psychrophilic fungus [73] that thrives at the body temperatures displayed by bats in torpor [74].

Despite intensive research, the origin of the pathogenic agent associated with this disease remains unknown and it is still not known why the disease appeared so suddenly [75]. The disease was first registered as a point-source epidemic at Howe's Cave, Albany, New York, in 2006 [71], since when it has spread westward at approx. 200–900 km annually [76]. Based on the 'novel pathogen hypothesis', Europe was initially thought to be the source of the agent which the following findings tending to suggest that WNS did indeed originate in Europe: (1) a single *P. destructans* genotype was identified in North American hibernating bats [77]; (2) the WNS fungal agent was also found in bats in European bats [78]; (3) no mass mortality events

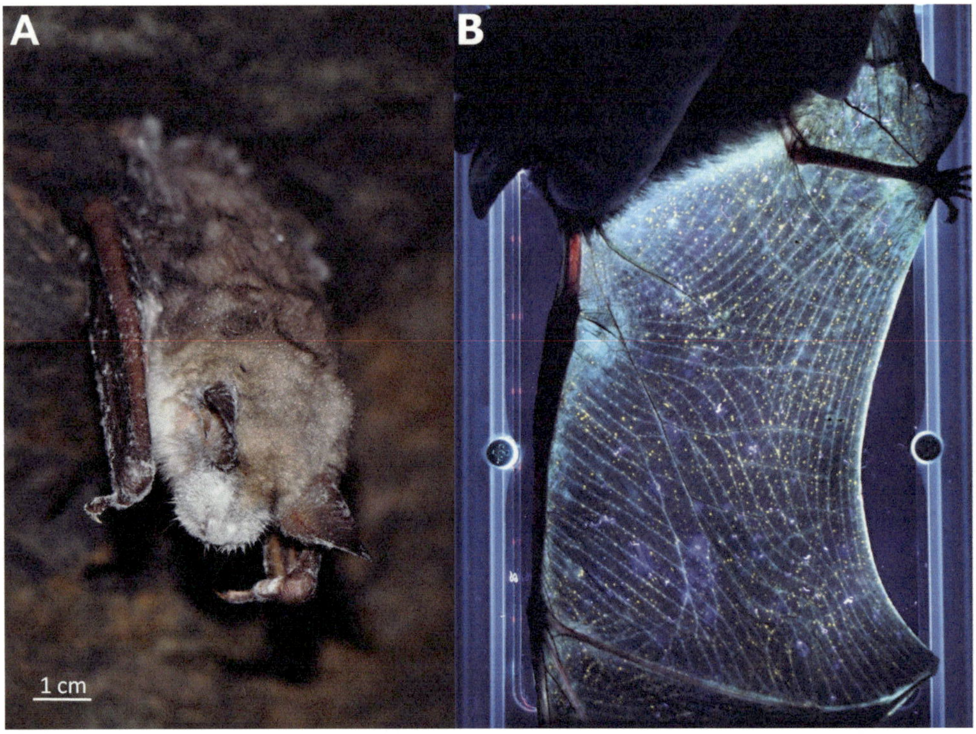

Figure 12. Skin infection with white-nose syndrome fungus. Greater mouse-eared bat (*Myotis myotis*) showing fungal growth on the muzzle, ears and forearm photographed in April 2016 (A). Transillumination of the wing membrane by ultraviolet light: a technique to visualize and detect wing membrane lesions based on yellow fluorescence of riboflavin produced by the white-nose syndrome agent *Pseudogymnoascus destructans* in the skin of affected bats (B). The scale is the same for both pictures. Photos by Jiri Pikula.

were reported in European hibernating bats harbouring the causative agent [79–81]; (4) inoculation with European fungus isolates induced WNS in the North American the little brown bat *Myotis lucifugus* [82]; and (5) European *P. destructans* isolates exhibited higher genetic diversity [83, 84]. However, recent findings of both the causative agent and WNS infection in Asia have refuted most of the hypothesis arguments and pushed the search for the pathogen source to non-European hibernacula [85, 86]. Likewise, the detection of bat WNS in western North America in March of 2016 does not fit the previously documented pattern of *P. destructans* geographic spread [76].

Diagnosis of WNS is based on identification of the fungal agent growing on bats using cultivation, morphological characteristics (e.g. crescent shaped conidia), and molecular assays [73, 87]. One of the most useful diagnostic methods is wing membrane transillumination with ultraviolet (UV) light, which reveals fluorescent lesions produced by the infection [88]. The method is non-lethal, can be used under field conditions and, in combination with photography, can be used to estimate infection intensity (**Figure 12B**) as it is highly sensitive and specific for WNS. Histopathology findings of typical cupping erosions (**Figure 13**) are the gold standard of WNS diagnosis [89, 90].

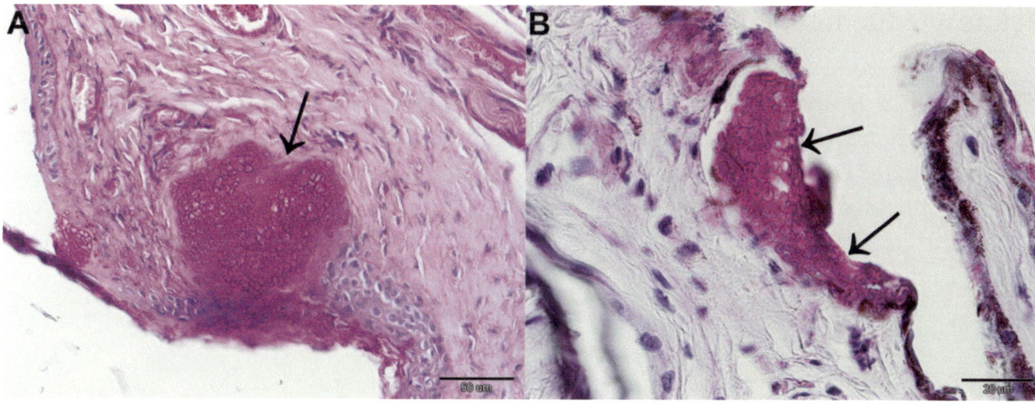

Figure 13. Microscopic appearance of white-nose syndrome infection. Thick-type skin of the muzzle extensively infected with *Pseudogymnoascus destructans* fungus (black arrow). Greater mouse-eared bat (*Myotis myotis*) sampled in a hibernaculum (Czech Republic) in May 2016 (A). Thin-type skin of the wing membrane with a white-nose syndrome pathognomonic cupping erosion that contains densely packed fungal hyphae (black arrows, B). The same bat specimen as in (A). Both skin sections stained with periodic acid-Schiff stain.

Surprisingly, the WNS fungal infection is restricted to the skin only, with no evidence of systemic fungal invasion in infected bats [71, 89]. Hence, bat mortality is thought to follow complex pathophysiological mechanisms, and a multi-stage WNS model has recently been proposed to explain the disease's progression [91]. Hibernating bats positive for WNS have been reported as displaying abnormal behaviour, higher arousal frequency from torpor, emaciation and fat depletion, dehydration, acidosis and electrolyte disbalance [82, 92–94]. The extent of wing pathology in infected bats appears to be directly related to mortality [95]. In general, Palearctic bats tend to have a lower disease intensity (measured as the percentage of wing membrane area affected by WNS lesions) than Nearctic bat species [96], possibly explaining the intercontinental differences in bat mortality.

Riboflavin or vitamin B2 is the main compound responsible for the distinctive orange-yellow fluorescence observed under UV light (**Figure 12B**) after invasive *P. destructans* growth has replaced living tissues. Pathogenic *P. destructans* strains produce considerably more riboflavin than non-pathogenic *Pseudogymnoascus* spp. strains. Importantly, high riboflavin concentrations accumulated in WNS skin lesions are toxic to cells under conditions typical for bat hibernation and euthermia. As such, riboflavin may act as a key virulence factor for WNS [96].

As *P. destructans* is a generalist pathogen, all bat species hibernating within contaminated caves may be at risk of infection [97]. However, adverse population-level effects depend on the species and appear to differ considerably between North America and Eurasia. Hibernating Palearctic bats appear to have evolved infection tolerance mechanisms to cope with the endemicity and extensive spatial distribution of virulent *P. destructans* in the Palearctic region [86]. These mechanisms include behavioural adaptations, such as specific patterns of hibernation and shelter selection [29] that ensure low pathogen impact. While our knowledge of this threat is growing, there are still numerous unanswered questions that require study, at the local and global levels.

Acknowledgements

We would like to thank all our colleagues and students from the Institute of Vertebrate Biology, Academy of Sciences of the Czech Republic Brno and Masaryk University in Brno, who participated in both field and laboratory research. We are especially grateful to Miroslav Kovařík for field assistance in the Moravian Karst. We are grateful to Kevin Roche for his correction and improvement of the English text. The authors received support from the Grant Agency of the Czech Republic (Grant no. 502/17-20286S).

Author details

Jan Zukal[1]*, Hana Berková[1], Hana Banďouchová[2], Veronika Kováčová[2] and Jiří Pikula[2]

*Address all correspondence to: zukal@ivb.cz

1 Institute of Vertebrate Biology, Academy of Sciences of the Czech Republic, Brno, Czech Republic

2 Department of Ecology and Diseases of Game, Fish and Bees, University of Veterinary and Pharmaceutical Sciences Brno, Brno, Czech Republic

References

[1] Kunz TH, Parsons S, editors. Ecological and Behavioral Methods for the Study of Bats. Baltimore: The Johns Hopkins University Press; 2009. p. 901

[2] Kunz TH, editor. Ecology of Bats. New York: Plenum Press; 1982. p. 425

[3] Findley JS. Bats: A Community Perspective. Cambridge: Cambridge University Press; 1993. p. 167

[4] Altringham JD. Bats: Biology and Behaviour. Oxford: Oxford University Press; 1996. p. 262

[5] Wilson DE, Reeder DAM, editors. Mammal Species of the World. A Taxonomic and Geographic Reference. Washington: Smithsonian Inst. Press; 1993. p. 1206

[6] Perry RW. A review of factors affecting cave climates for hibernating bats in temperate North America. Environmental Research. 2013;**21**:28-39

[7] Mitchell-Jones A J, Bihari Z, Masing M, Rodrigues L. Protecting and Managing Underground Sites for Bats. Bonn, Germany: EUROBATS Publication Series No. 2. UNEP/EUROBATS Secretariat; 2007. p. 38

[8] Erkert HG. Ecological aspects of bat activity rhythms. In: Kunz TH, editor. Ecology of Bats. New York: Plenum Press; 1982. pp. 201-242

[9] Řehák Z, Zukal J, Kovařík M. Long- and short-term changes in the bat community of the Kateřinská cave (Moravian Karst)—a fundamental assessment. Folia Zoologica. 1994;**43**:425-436

[10] Zima J, Kovařík M, Gaisler J, Řehák Z, Zukal J. Dynamics of the number of bats hibernating in the Moravian Karst in 1983 to 1992. Folia Zoologica. 1994;**43**:109-119

[11] Degn HJ, Andersen BB, Baagoe H. Automatic registration of bat activity through the year at Monsted Limestone Mine, Denmark. Zeitschrift für Säugetierkunde. 1995;**60**:129-135

[12] Harrje C. Etho-ökologische Untersuchungen an winterschlafenden Wasserfleder-mäusen (*Myotis daubentoni*). Nyctalus (N.F.), Berlin. 1999;**7**:78-86

[13] Simon M, Kugelschafter K. Die Ansprüche der Zwergfledermaus (*Pipistrellus pipistrellus*) an ihr Winterquartier. – Nyctalus (N.F.), Berlin. 1999;**7**:102-111

[14] Parsons KN, Jones G, Greenaway F. Swarming activity of temperate zone microchiropteran bats: Effects of season, time of night and weather conditions. Journal of Zoology, London. 2003;**261**:257-264

[15] Rivers NM, Butlin RK, Altringham JD. Autumn swarming behaviour of Natterer's bats in the UK: Population size, catchment area and dispersal. Biological Conservation. 2006;**27**:215-226

[16] Ransome RD. The Natural History of Hibernating Bats. London: Christopher Helm; 1990. p. 256

[17] Baroň I. Seasonal changes in flight activity of lesser horseshoe bat (*Rhinolophus hipposideros*) in cave entrance [thesis]. Brno: PřF MU 2000. p. 53

[18] Park JK, Jones G, Ransome RD. Winter activity of a population of greater horseshoe bats (*Rhinolophus ferrumequinum*). Journal of Zoology, London. 1999;**248**:419-427

[19] Glover AM, Altringham JD. Cave selection and use by swarming bat species. Biological Conservation. 2008;**141**:1493-1504

[20] Parsons S, Jones G. Acoustic identification of twelve species of echolocating bat by discriminant function analysis and artificial neural networks. Journal of Experimental Biology. 2000;**203**:2641-2656

[21] Berková H, Zukal J. Flight activity of bats at the entrance of a natural cave. Acta Chiropterologica. 2006;**8**:187-195

[22] Berková H, Zukal J. Cave visitation by temperate zone bats: Effects of climatic factors. Journal of Zoology, London. 2010;**280**:387-395

[23] Speakman JR, Racey PA. Hibernal ecology of the pipistrelle bat: Energy expenditure, water requirements and mass loss, implications for survival and the function of emergence flights. Journal of Animal Ecology. 1989;**58**:797-813

[24] Thomas DW. Hibernating bats are sensitive to nontactile human disturbance. Journal of Mammalogy. 1995;**76**:940-946

[25] Boyles JG, Dunbar MB, Whitaker Jr. JO. A review of activity following arousal in winter in North American vespertilionid bats. Mammal Review. 2006;**36**:267-280

[26] Thomas DW. Lack of evidence for a biological alarm clock in bats (*Myotis* spp.) hibernating under natural conditions. Canadian Journal of Zoology. 1993;**71**:1-3

[27] Nagel A, Nagel R. Nutzung eines Untertagequartieres durch die Kleine Hufeisennase (*Rhinolophus hipposideros*). Tagungsband: "Zur Situation der Hufeisennasen in Europa" Nebra, den 26. - 28. Mai 1995, Arbeitskreis Fledermäuse Sachsen-Anhalt e.V. 1997:97-108

[28] Hope PR, Jones G. Warming up for dinner: Torpor and arousal in hibernating Natterer's bats (*Myotis nattereri*) studied by radio telemetry. Journal of Comparative Physiology B. 2012;**182**:569-578

[29] Zukal J, Berková H, Madaraszová J. Flying or sleeping: Influence of WNS on flight activity of bats during deep hibernation. Folia Zoologica. 2016;**65**:46-51

[30] Zukal J, Berková H, Řehák Z. Activity and shelter selection by *Myotis myotis* and *Rhinolophus hipposideros* hibernating in Kateřinská cave (Czech Republic). Mammalian Biology. 2005;**70**:271-281

[31] Ciechanowski M, Zajac T, Bilas A, Dunajski R. Spatiotemporal variation in activity of bat species differing in hunting tactics: Effects of weather, moonlight, food abundance, and structural clutter. Canadian Journal of Zoology. 2007;**85**:1249-1263

[32] Dunbar MB, Whitaker Jr. JO, Robbins LW. Winter feeding by bats in Missouri. Acta Chiropterologica. 2007;**9**:305-310

[33] Skiba R. Bestandsentwicklung und Verhalten von Fledermäusen in einem Stollen des Westharzes. Myotis. 1987;**25**:95-103

[34] Bauerová Z, Zima J. Seasonal changes in visits to a cave by bats. Folia Zoologica. 1988;**37**:97-111

[35] Whitaker Jr. JO, Rissler LJ. Seasonal activity of bats at Copperhead cave. Proceedings of the Indiana Academy of Science. 1992;**101**:127-134

[36] Anthony ELP, Stack MH, Kunz TH. Night roosting and the nocturnal time budget of the little brown bat, *Myotis lucifugus*: Effects of reproductive status, prey density, and environmental conditions. Oecologia. 1981;**51**:151-156

[37] Lučan R, Hanák V. Population structure of Daubenton's bats is responding to microclimate of anthropogenic roosts. Biologia. 2011;**66**:690-695

[38] Horáček I, Zima J. Net-revealed cave visitation and cave-dwelling on European bats. Folia Zoologica. 1978;**27**:135-148

[39] Horáček I, Uhrin M, editors. A tribute to bats. Kostelec nad Černými lesy: Lesnická práce; 2010. p. 314.

[40] Uhrin M, Benda P, Obuch J, Urban P. Changes in abundance of hibernating bats in central Slovakia (1992-2009). Biologia. 2010;**65**:349-361

[41] Kolenati FA.Monographie der europäischen Chiropteren. Jarhesh. natw. Sect. k.k. Mähr.-Schl. Gesellsch. Förd. Brünn: Ackerbaues; 1860. p. 156

[42] Gaisler J, Řehák Z, Zukal J. Bat research in the Moravian Karst: History and present state. Vespertilio. 2006;**9-10**:75-85

[43] Zukal J, Kovařík M, Řehák Z, Berková H. Numbers of bats hibernating in two caves in the northern part of Moravian Karst and their long-term changes. Vespertilio. 2001;**5**:321-328

[44] Ewing WG, Studier EH, O'Farrell MJ. Autumn fat deposition and gross body composition in three species of *Myotis*. Comparative Biochemistry and Physiology. 1970;**36**:119-129

[45] Avery MI. Winter activity of pipistrelle bats. Journal of Animal Ecology. 1985;**54**:721-738

[46] Brack Jr. V, Twente JW. The duration of the period of hibernation of three species of vespertilionid bats. I. Field studies. Canadian Journal of Zoology. 1985;**63**:2952-2954

[47] French AR. Allometries of the durations of torpid and euthermic intervals during mammalian hibernation: A test of the theory of metabolic control of the timing of changes in body temperature. Journal of Comparative Physiology B. 1985;**156**:13-19

[48] Geiser F, Ruf T. Hibernation versus daily torpor in mammals and birds: Physiological variables and classification of torpor patterns. Physiological Zoology. 1995;**68**:935-966

[49] Zukal J, Řehák Z, Kovařík M. Bats of Sloupsko-šošůvské cave (Moravian Karst, Central Moravia). Lynx (Praha) n.s. 2003;**34**:205-220

[50] Harmata W. Dynamika fenologiczna hibernacji podkowców małych, *Rhinolophus hipposideros* (Bechst.) (Chiroptera, Rhinolophidae). Studia Chiropterologica. 2000;**1**:13-28

[51] Nagel A, Nagel R. Bestandsentwicklung winterschlafender Fledermäuse auf der Schwäbischen Alb bis zum Winter 1987/88 und ihr Schutz. Mitt. Verb. dt. Höhlen- und Karstforsch. 1989;**35**:17-23

[52] Ransome RD. The distribution of the Greater horse-shoe bat, *Rhinolophus ferrumequinum*, during hibernation, in relation to environmental factors. Journal of Zoology, London. 1968;**154**:77-112

[53] Nagel A, Nagel R. Remarks of the pattern of optimal ambient temperatures in hibernating bats. Myotis. 1991;**29**:109-114

[54] Postawa T. A cave microclimate as modelled by external climatic conditions and its effect on a hibernating bat assemblage: A case of the "Pod Sokolą Górą" cave. Proceedings of the VIIIth EBRS. 2000;**1**:199-217

[55] Brosset A, Poillet A. Structure d'une population hibernante de grands rhinolophes *Rhinolophus ferrumequinum* dans l'est de la France. Mammalia. 1985;**49**:221-233

[56] Hanzal V, Průcha M. Annual course of cave visitation by bats (Mammalia: Chiroptera) in the Bohemian karst (Czech Republic). Acta Societatis Zoologicae Bohemoslovacae. 1996;**60**:25-30

[57] Fuszara E, Kowalski M, Lesinski G, Cygan JP. Hibernation of bats in underground shelters of central and northeastern Poland. Bonner Zoologische Beiträge. 1996;**46**:349-358

[58] Dunbar MB, Brigham RM. Thermoregulatory variation among populations of bats along a latitudinal gradient. Journal of Comparative Physiology. 2010;**180**:885-893

[59] Pörter HO, Bennett AF, Bozinovic F, Clarke A, Lardies MA, Lucassen M, Pelster B, Schiemer F, Stillman JH. Tradeoffs in thermal adaptation: The need for a molecular to ecological integration. Physiological and Biochemical Zoology. 2006;**79**:296-313

[60] Gaisler J, Hanák V, Hanzal V, Jarský V. Results of bat banding in the Czech and Slovak Republics, 1948-2000. Vespertilio. 2003;**7**:3-61

[61] Nagel A, Nagel, R. How do bats choose optimal temperatures for hibernation? Comparative Biochemistry and Physiology. 1991;**99A**:109-114

[62] Webb PI, Speakman JR, Racey PA. How hot is a hibernaculum? A review of the temperatures at which bats hibernate. Canadian Journal of Zoology. 1995;**74**:761-765

[63] Boyles JG, Dunbar MB, Storm JS, Brack Jr. V. Energy availability influences microclimate selection of hibernating bats. Journal of Experimental Biology. 2007;**210**:4345-4350

[64] Kuipers B, Daan S. Internal migration of hibernating bats: Response to seasonal variation in cave microclimate. Bijdragen tot de Dierkunde, Amsterdam. 1970;**40**:51-55

[65] Luis AD, Hudson OJ. Hibernation patterns in mammals: A role for bacterial growth? Functional Ecology. 2006;**20**:471-477

[66] Humphries MM, Thomas DW, Kramer D. The role of energy availability in mammalian hibernation: A cost-benefit approach. Physiological and Biochemical Zoology. 2003;**76**:165-179

[67] Clawson RL, Laval RK, Laval ML, Caire W. Clustering behavior of hibernating *Myotis sodalis* in Missouri. Journal of Mammalogy. 1980;**61**:245-253

[68] Fujita J. Cold shock response in mammalian cells. Journal of Molecular Microbiology and Biotechnology. 1999;**1**:243-255

[69] Bartonička T, Bandouchova H, Berková H, Blažek J, Lučan R, Horáček I, Martínková N, Pikula J, Řehák Z, Zukal J. Deeply torpid bats can change position without elevation of body temperature. Journal of Thermal Biology. 2017;**63**:119-123

[70] Boyles JG, Cryan PM, McCracken GF, Kunz TH. Economic importance of bats in agriculture. Science. 2011;**332**:41-42

[71] Blehert DS, Hicks AC, Behr M, Meteyer CU, Berlowski-Zier BM, Buckles EL, Coleman JTH, Darling SR, Gargas A, Niver R, Okoniewski JC, Rudd RJ, Stone WB. Bat white-nose syndrome: An emerging fungal pathogen? Science. 2009;**323**:227

[72] Lorch JM, Meteyer CU, Behr MJ, Boyles JG, Cryan PM, Hicks AC, Ballmann AE, Coleman JTH, Redell DN, Reeder DM, Blehert DS. Experimental infection of bats with *Geomyces destructans* causes white-nose syndrome. Nature. 2011;**480**:376-378

[73] Gargas A, Trest MT, Christensen M, Volk TJ, Blehert DS. *Geomyces destructans* sp. nov. associated with bat white-nose syndrome. Mycotaxon. 2009;**108**:147-154

[74] Verant ML, Boyles JG, Waldrep Jr. W, Wibbelt G, Blehert DS. Temperature-dependent growth of *Geomyces destructans,* the fungus that causes bat white-nose syndrome. PLoS ONE. 2012;**7**, e46280

[75] Coleman JTH, Reichard JD. Bat white-nose syndrome in 2014: A brief assessment seven years after discovery of a virulent fungal pathogen in North America. Outlooks on Pest Management. 2014;**25**:374-377

[76] Lorch JM, Palmer JM, Lindner DL, Ballmann AE, George KG, Griffin K, Knowles S, Huckabee JR, Haman KH, Anderson CD, Becker PA, Buchanan JB, Foster JT, Blehert DS. First detection of bat white-nose syndrome in western North America. mSphere. 2016;**1**:00148-16

[77] Sunanda SR, Li X, Rudd RJ, Okoniewski JC, Xu J, Chaturvedi S, Chaturvedi V. Clonal genotype of *Geomyces destructans* among bats with white nose syndrome, New York, USA. Emerging Infectious Disease Journal. 2011;**17**:1273-1276

[78] Puechmaille SJ, Verdeyroux P, Fuller H, Ar Gouilh M, Bekaert M, Teeling EC. White-nose syndrome fungus (*Geomyces destructans*) in bat, France. Emerging Infectious Diseases. 2010;**16**:290-293

[79] Martínková N, Bačkor P, Bartonička T, Blažková P, Červený J, Falteisek L, Gaisler J, Hanzal V, Horáček D, Hubálek Z, Jahelková H, Kolařík M, Korytár L, Kubátová A, Lehotská B, Lehotský R, Lučan R, Májek O, Matějů J, Řehák Z, Šafář J, Tájek P, Tkadlec E, Uhrin M, Wagner J, Weinfurtová D, Zima J, Zukal J, Horáček I. Increasing incidence of *Geomyces destructans* fungus in bats from the Czech Republic and Slovakia. PLoS ONE. 2010;**5**:e13853

[80] Puechmaille SJ, Wibbelt G, Korn V, Fuller H, Forget F, Mühldorfer K, Kurth A, Bogdanowicz W, Borel C, Bosch T, Cherezy T, Drebet M, Görföl T, Haarsma AJ, Herhaus F, Hallart G, Hammer M, Jungmann C, Le Bris Y, Lutsar L, Masing M, Mulkens B, Passior K, Starrach M, Wojtaszewski A, Zöphel U, Teeling EC. Pan-European distribution of white-nose syndrome fungus (*Geomyces destructans*) not associated with mass mortality. PLoS ONE. 2011;**6**:e19167

[81] Wibbelt G, Kurth A, Hellmann D, Weishaar M, Barlow A, Veith M, Pruger J, Gorfol T, Grosche L, Bontadina F, Zöphel U, Seidl HP, Cryan PM, Blehert DS. White-nose syndrome fungus (*Geomyces destructans*) in bats, Europe. Emerging Infectious Diseases. 2010;**16**:1237-1243

[82] Warnecke L, Turner JM, Bollinger TK, Lorch JM, Misra V, Cryan PM, Wibbelt G, Blehert DS, Willis CKR. Inoculation of bats with European *Geomyces destructans* supports the novel pathogen hypothesis for the origin of white-nose syndrome. Proceedings of the National Academy of Sciences. 2012;**109**:6999-7003

[83] Leopardi S, Blake D, Puechmaille SJ. White-nose syndrome fungus introduced from Europe to North America. Current Biology. 2015;**25**:R217–R219

[84] Palmer JM, Kubatova A, Novakova A, Minnis AM, Kolarik M, Lindner DL. Molecular characterization of a heterothallic mating system in *Pseudogymnoascus destructans*, the fungus causing white-nose syndrome of bats. G3: Genes|Genomes|Genetics. 2014;**4**: 1755-1763

[85] Hoyt JR, Sun K, Parise KL, Lu G, Langwig KE, Jiang T, Yang S, Frick WF, Kilpatrick AM, Foster JT, Feng J. Widespread bat white-nose syndrome fungus, Northeastern China. Emerging Infectious Disease Journal. 2016;**22**:140-142

[86] Zukal, J, Banďouchová H, Brichta J, Čmoková A, Jaroň KS, Kolařík M, Kováčová V, Kubátová A, Nováková A, Orlov O, Pikula J, Presetnik P, Šuba J, Zahradníková A, Martínková N. White-nose syndrome without borders: *Pseudogymnoascus destructans* infection tolerated in Europe and Palearctic Asia but not in North America. Scientific Reports. 2016;**6**:19829

[87] Shuey MM, Drees KP, Lindner DL, Keim P, Foster JT. Highly sensitive quantitative PCR for the detection and differentiation of *Pseudogymnoascus destructans* and other *Pseudogymnoascus* species. Applied and Environmental Microbiology. 2014;**80**:1726-1731

[88] Turner GG, Meteyer CU, Barton H, Gumbs JF, Reeder DA, Overton B, Bandouchova H, Bartonička T, Martínková N, Pikula J, Zukal J, Blehert DS. Nonlethal screening of bat-wing skin with the use of ultraviolet fluorescence to detect lesions indicative of white-nose syndrome. Journal of Wildlife Diseases. 2014;**50**:566-573

[89] Meteyer CU, Buckles EL, Blehert DS, Hicks AC, Green DE, Shearn-Bochsler V, Thomas NJ, Gargas A, Behr MJ. Histopathologic criteria to confirm white-nose syndrome in bats. Journal of Veterinary Diagnostic Investigation: Official Publication of the American Association of Veterinary Laboratory Diagnosticians, Inc. 2009;**21**:411-414

[90] Pikula J, Banďouchová H, Novotný L, Meteyer CU, Zukal J, Irwin NR, Zima J, Martínková N. Histopathology confirms white-nose syndrome in bats in Europe. Journal of Wildlife Diseases. 2012;**48**:207-211

[91] Verant ML, Meteyer CU, Speakman JR, Cryan PM, Lorch JM, Blehert DS. White-nose syndrome initiates a cascade of physiologic disturbances in the hibernating bat host. BMC Physiology. 2014;**14**:10

[92] Cryan PM, Meteyer CU, Blehert DS, Lorch JM, Reeder DM, Turner GG, Webb J, Behr M, Verant M, Russell RE, Castle KT. Electrolyte depletion in white-nose Syndrome Bats. Journal of Wildlife Diseases. 2013;**49**:398-402

[93] Reeder DM, Frank CL, Turner GG, Meteyer CU, Kurta A, Britzke ER, Vodzak ME, Darling SR, Stihler CW, Hicks AC, Jacob R, Grieneisen LE, Brownlee SA, Muller LK, Blehert DS. Frequent arousal from hibernation linked to severity of infection and mortality in bats with white-nose syndrome. PLoS ONE. 2012;**7**:e38920

[94] Warnecke L, Turner JM, Bollinger TK, Misra V, Cryan PM, Blehert DS, Wibbelt G, Willis CKR. Pathophysiology of white-nose syndrome in bats: A mechanistic model linking wing damage to mortality. Biology Letters. 2013;**9**:20130177

[95] Cryan P, Meteyer C, Boyles J, Blehert D. Wing pathology of white-nose syndrome in bats suggests life-threatening disruption of physiology. BMC Biology. 2010;**8**:135

[96] Flieger, M. et al. Vitamin B2 as a virulence factor in *Pseudogymnoascus destructans* skin infection. Scientific Reports. 2016;**6**:33200

[97] Zukal J, Banďouchová H, Bartonička T, Berková H, Brack V, Brichta J, Dolinay M, Jaroň KS, Kováčová V, Kovařík M, Martínková N, Ondráček K, Řehák Z, Turner GG, Pikula J. White-nose syndrome fungus: A generalist pathogen of hibernating bats. PLoS ONE. 2014;**9**:e97224

Chapter 4

Important Caves in Turkish Thrace for Bats: Dupnisa Cave System and Koyunbaba Cave

Serbülent Paksuz

Additional information is available at the end of the chapter

http://dx.doi.org/10.5772/intechopen.68836

Abstract

Today, caves and some bat populations are declining worldwide because of anthropogenic pressures such as habitat use, tourism, quarrying, and guano mining. The protection of caves is very essential for bat conservation programs because some caves are indispensable and specific living areas for many bat populations. In this chapter, the species composition, colony structure, seasonal population dynamics, roosting habits, and roosting requirements of the bats in Dupnisa Cave System and Koyunbaba Cave in Turkish Thrace are discussed. Dupnisa Cave System and Koyunbaba Cave, which have different roost characteristics and microclimates, are alternative to each other in terms of the season. Therefore, they are inhabited by different bat species for different purposes at different levels according to weather conditions changing throughout the year. Dupnisa Cave System is mainly used by 18 bat species for hibernating, whereas Koyunbaba Cave is mainly used by 11 bat species for breeding and nursing. Due to different roost characteristics and microclimatic conditions, Dupnisa Cave System and Koyunbaba Cave are the most important underground habitats for bat populations in Turkish Thrace. Therefore, the protection of these caves is very important for the future of bat populations in the region.

Keywords: bats, caves, conservation, Dupnisa Cave System, Koyunbaba Cave, Turkish Thrace

1. Introduction

This chapter summarizes general information about cave ecosystems and bats. In the first part, the characteristics and importance of caves and bats will be elaborated for interested readers who are not experts in these subjects. In the following sections, geomorphology and biogeography of Turkey and Turkish Thrace are discussed in detail, and the characteristics that make the region important are also underlined. In conclusion of the chapter, more detailed data about Dupnisa Cave System and Koyunbaba Cave in Turkish Thrace will be presented, and the importance of these caves for the future of the bats in the region will be revealed. This chapter presents a summary of related previous studies. The aim is to raise awareness about the protection of bats and caves rather than be a reference for similar topics.

1.1. Cave ecosystems and their characteristics

In simple terms, cave is defined as an underground hollow that is large enough for a person to enter; however, cave has various definitions in terminology. Caves are formed as a result of the dissolution of carbonated and sulfated rocks with a series of physiochemical processes by underground waters. Caves contain the past and current data of the geological, geomorphological, hydrological, and ecological characteristics of the region where they are located [1]. Caves are not only the underground hollows formed in the rocks but also the unique habitats that host many organisms and natural values. Caves are important ecosystems in terms of both cave-dwelling species, and geological, historical, anthropological, archaeological, ecological, and cultural values.

Caves are very specific and susceptible ecosystems, because they have some extreme features that force life such as darkness, low temperature, high humidity, and limited nutrients [2–4]. Because of their darkness, caves are devoid of plants, which are the main producers of the food chain. For this reason, all the organic materials are brought from the outside into the cave in various ways. The main factors that provide nutrient entry into the caves are the air flows and the water currents in the caves. Besides these, guano, the main energy source in the growth of many organisms, is the most important nutrient in caves [5]. Lack of light, low temperature, high humidity, and the restriction of nutrients can be seen as important limiting factors in many caves [3, 4]. In the caves, living organisms are resistant or adapted to the scarcity of nutrients and other limiting factors. The essential condition for organisms to survive in the cave is energy saving. For this reason, the cave organisms are as small as possible and they move as little as possible.

The most important difference between underground and terrestrial habitats is the stable environmental conditions of caves. Unlike terrestrial conditions, caves have an almost constant temperature and humidity throughout the year. All these differences and characteristics make cave habitats unique and very sensitive. Any changes in the conditions of the cave ecosystems threaten the lives of the cave inhabitants. Despite challenging living conditions such as darkness and limited nutrient sources, caves accommodate many species [3, 4]. Because of these negative conditions in the cave, the cave dwellers have developed various ways of adaptations to survive. For example, the sense of sight of organisms adapted to life in caves has either weakened or completely disappeared in the evolutionary process. On the other hand, most of these species have developed their antennas longer than their relatives outside. Similarly, some species that adapted to the cave life usually appear white or transparent because they lack pigment.

http://dx.doi.org/10.5772/intechopen.68836

Caves are initially creepy, but the mystery of the darkness and the past in caves is the most important features that attract people into caves. And also, in the caves formed as a result of a very slow and long process, stalactites, stalagmites, and travertines come in various formations, which attract people visually. In addition, underground waters, rivers, and lakes within caves are another beauty. So, it is possible to see that cave tourism, which is called as alternative tourism, has improved all over the world. The visual, sportive, religious, historical, and cultural characteristics of the caves provide a potential for tourism. However, it is also known that caves are used for different purposes such as shelter, barn, and storage. For this reason, it is inevitable that underground habitats, which are natural habitats of organisms, are damaged by human pressures over time.

Underground habitats around the world are under constant threat due to human pressures such as cave tourism, treasure hunting, guano mining, stone quarries, and water dams. The use of caves by humans can cause significant harmful effects on caves [6–8]. It should not be forgotten that these natural values, gradually formed in millions of years, will not come back again in a short time. For this reason, the identification and investigation of important cave habitats are essential for the future of these habitats and organisms living there.

1.2. Bats and their characteristics

Bats (Chiroptera) are among the most diverse and widely distributed groups of mammals and can be found in most continents. Bats consist of greater than 1300 species worldwide and comprise approximately one quarter of mammalian species richness [9–11]. The order Chiroptera is subdivided into the Megachiroptera and the Microchiroptera. The Megachiroptera is represented by only one family, whereas the Microchiroptera comprise 17 families. Megabats are also called as fruit bats. Fruit bats mostly roost in trees and shrubs. Megabats are frugivorous and nectarivorous, and they eat fruit or lick nectar from flowers. Microbats are called as insectivorous, and they mostly feed on insects, and use echolocation. Insectivorous bats use a wide variety of roosts such as caves, trees, and manmade structures.

Bats are the only flying mammals and they have a wide range of feeding and roosting habits, social behaviors, and reproductive strategies. Flying ability made bats become one of the most common groups of mammals [6]. Most of bat species use echolocation to navigate and forage. Echolocation allowing bats to see their surroundings by sound is actually an adaptation. Bats emit calls out to the environment through their nostrils and mouth, and then they find their direction by listening to the echoes of the calls returning from the objects around them. Thus, bats get information about the location, position, and shape of objects in the surroundings by processing these echoes [12, 13]. When food quantity decreases, bats have to make a choice: to hibernate or to migrate. Hibernation involves a reduction in metabolic rate that allows bats to survive for a long time without food [14]. Many species of bats migrate between summer and winter roosts. Some migrations are made to reach for more abundant food sources in warmer locations, while others are made to obtain roosts that have optimal microclimate for hibernating in winter or raising offspring in summer [14, 15].

Bats have acquired a wide variety of characteristics in parallel with the lifestyle they are adapted. Although the patterns of feeding, roosting, and reproduction of the bats differ between species, some basic adaptations are similar. Most bat species are nocturnal, and they forage during the nights. Bats feed with a wide variety of foods such as insect, fruit, flower nectar, small vertebrate, and blood [12, 15]. In the daylight, bats pass into torpor in roosts such as foliage, caves, hollows of trees, rock crevices, and various manmade structures. The most prominent predators of bats are owls, hawks, snakes, raccoons, and martens [15]. Female bats have some strategies that they can control the timing of their pregnancy to give birth at an appropriate time [13]. Females generally give birth to one offspring every year, and they nurse their young until they get mature to feed on their own [15].

Bats that play a key role in many ecosystems are an important group of mammals with species diversity, abundance, and distribution all over the world. Bats have important ecological roles that are critical for human and ecosystem health, including the pollination and dispersal of many plants, and the control of insects [16, 17]. Bats are primary predators of insects that are harmful for human health and agriculture. Thus, bats play an important role in human health and biological pest control by balancing harmful insect and microorganism populations [15]. Guano mined from caves is provided by bats and used as a natural fertilizer on agricultural crops. Guano is a primary source of nutrition that allows the development of a great diversity of organisms such as arthropod, fungi, bacteria, and lichen.

Bats are one of the most sensitive mammal groups to varying weather conditions with their peculiarities in physiology, thermoregulation, and life cycle. Bats prefer different roosts depending on the season and spend more than half of their lives in roost. Most species of bats have specific requirements in terms of roost conditions such as microclimate and environmental stability [18]. Roosts protect bats from bad weather conditions and their predators, so roost selection is essential to sustain the life of the bats [13]. Therefore, the presence of suitable roost is an important factor affecting the social structure and the distribution of bats [18–20]. The identification of these roosts is essential for the protection of bats. Bats have been known as cave dwellers for a long time, because caves that provide stable environmental conditions and protection throughout the year are pretty suitable roosts for bats. All over the world, caves and mines are used as shelter by large bat populations, but particularly sensitive species are highly dependent on only a few of these underground shelters [9, 18, 21]. However, most caves are used only by a small number of bats. Only a few caves provide convenient conditions to host thousands of bats, especially during hibernating and breeding periods. These caves are vital to bat populations [9, 18, 20, 22].

Microclimatic conditions are often particularly important for habitat quality of bats [23]. Therefore, the seasonal use of roosts by bats is profoundly associated with its microclimatic conditions and these conditions form the assembling patterns of bats. The particular roost requirements of bat species restrict the permanent shelters that are used by bats. Thus, the environmental stability and protection provided by caves make them highly suitable roost for bats throughout the year. Most bat species make use of caves as roost for various purposes during some phases of their annual cycle [22].

Microclimatic requirements of bats show seasonal variations according to their annual life cycle. Therefore, the availability of a cave by bats is increased by providing alternative microclimatic conditions compared to season. Caves may serve as one of the most adequate roosting sites for bats because of the relatively stable microclimatic characteristics. Therefore, natural caves and artificial underground sites are widely used as roosting sites by bats and may be occupied by large breeding and/or hibernating populations [9]. The distribution and availability of suitable roosting sites is a limited resource for some cave-dwelling bat species [20]. Besides, cave-dwelling bats are mostly threatened by visitors and human activities.

The distribution areas of animals depend on their biological requirements and environmental conditions. Because of the variety of such conditions, the caves, which are potential roosts for bats, can serve as a good model for studying the relationship between the regional settlement and the structure of the environment [24]. Bats have to find special periodic roosts according to the requirements of hibernation or breeding, so knowing the roosting requirements of bats is vital for conservation and management works. Identifying and surveying these sites will help to understand the habitat requirements of species [25].

Bats form the largest mammalian assemblages on earth, and one place may be a shelter for a significant portion of the total populations of some species [26]. Such places have a great importance for bats and may be limiting for their population size and distribution. Most bat species rely on underground shelters to maintain their lives. Cave disturbance and destruction is one of the biggest problems that cause the decline of bat populations all over the world, so the identification and protection of such important sites has vital importance for the future of bats [8].

1.3. Biogeography and geomorphology of Turkey

When compared to the regional countries, Turkey has a special biogeography due to its large surface area, different geomorphologic structure, different climatic conditions, and transition position between the continents [27]. As a result of these features, Turkey has a rich biodiversity and a variety of ecosystems that almost a continent can have. The geomorphological structure of Turkey is one of the main factors in having various biogeographic areas [27].

The soluble karstic rocks such as limestone, gypsum, and dolomite, which are suitable for cave formation, constitute approximately 40% of Turkey's surface area [28]. One of the most important formations in the karstic areas of Turkey is caves. Because of the insufficiency of cave explorations, not all of the caves in Turkey have yet been examined; it is not possible to give a definite figure about the number of caves in our country. However, it is estimated that more than 20,000 caves can be found in Turkey according to the size of the area covered by the soluble rocks and the ratio of the number of caves detected in these areas [28]. However, many of the caves in Turkey are still not identified. In parallel, research on caves in Turkey is extremely inadequate. The research carried out until now is mostly focused on the caves in the touristic areas. Approximately 3000 of these caves were surveyed by caving associations, clubs, communities, and organizations. Until today, 31 caves in Turkey have been opened as tourist attractions [29].

1.4. Biogeography and geomorphology of Turkish Thrace

Thrace is a geographic region located in southeastern Europe within the territories of Greece, Bulgaria, and Turkey. Thrace lies in the northeastern Greece, the southeastern Bulgaria, and the European part of Turkey (**Figure 1**). Turkish Thrace, which constitutes about 3% of Turkey's surface, is not very high (average altitude of 180 m) and only 15% of the region is covered by forest in north and south. When evaluated in terms of geological, geographical, and tectonic features, it is seen that the Thrace region is different from the other regions of Turkey. Karst formation in Turkish Thrace is characterized by Eocene limestone, but a few caves developed within the marbles of the Paleozoic metamorphic basement, which is limited in the region [30].

Thrace is one of the major biogeographic zones in Turkey and, due to its karst formation, more than 50 caves have been formed in the region. Most of these caves, which form horizontally, are fossil-typed caves that have completed development. The caves in Turkish Thrace are concentrated at elevations between 150 and 200 m in Yıldız Mountains. The two longest caves explored were İkigöz Cave (4816 m) and Dupnisa Cave System (2720 m) in length with altitudes of 70 and 345 m, respectively [30, 31].

Despite the presence of thousands of caves in Turkey, there are few studies of the cave fauna, and most are old and based on the limited sampling. However, in recent years, the caves in Turkish Thrace have been slightly more surveyed in terms of the bats. Bilgin [32] has examined some of the caves in the region and provided information on the summer population sizes and habitat preferences of the bat species. Furman and Özgül [33, 34] have investigated the population sizes of bat species in many caves in the region and also stated that they should be protected by identifying important underground shelters in the region. In particular, two caves in the region, Dupnisa Cave System and Koyunbaba Cave, have been investigated in

Figure 1. The location map of Thrace, and the positions of Dupnisa Cave System and Koyunbaba Cave in Turkish Thrace.

detail by the author based on regular and long-term monitoring (**Figure 1**) [35–38]. In recent years, many studies have been conducted throughout Turkey to determine other underground habitats that are important for bat populations. These studies provide important contributions to the identification and conservation of the bat populations and roosts in Turkey [39, 40].

Due to its location and different climatic zones, Turkey has become a homeland and shelter for many species affected by geological and climatic changes in the past. For this reason, Turkey, which has a rich biodiversity, is also very rich in terms of bat species. Turkey is the country with the highest bat diversity in the region with 39 species identified so far. In Turkish Thrace, 27 of these species have been recorded. Dupnisa Cave System, Koyunbaba Cave, and Kocakuyu Cave are the most important shelters for bats in the Thrace region of Turkey [32–38].

Besides the lack of suitable shelters, today's rapidly growing urbanization, population growth, and the increased interest of humans intended for nature are causing damage and gradual decline of natural habitats of bats, and thus bat populations are at risk of extinction. Cave-dwelling bats are mostly threatened by visitors and human activities. Treasure hunting, quarrying, and cave tourism are the main threats to caves in Turkish Thrace. Knowing and protecting of roosts are essential to bat conservation. The use of caves by bats is shaped by microclimatic, morphological, and ecological conditions provided by the caves, because species have specific requirements for roosts in terms of shelter. Roost selection of bats differs in particular parts of caves and changes according to their annual life cycle. Therefore, knowledge about roost requirements and roost switching of the bat species in Turkish Thrace is required to any plan of the protection of the bats.

2. Methodology and study area

In this study, Dupnisa Cave System and Koyunbaba Cave in Turkish Thrace were investigated based on a long-term monitoring. The data given in this chapter belong to the years 2002–2010 for Dupnisa Cave System and 2007–2008 for Koyunbaba Cave. The roosts used by bats in both the caves were determined. In each survey, the species composition, colony structure, seasonal population numbers, and roosting habits of the bats were recorded. Roost temperature and humidity were regularly measured in each survey. Collected data were evaluated in two periods as winter/hibernation (November–March) and summer/nursery (April–October). Species identifications were done by following the identification keys [41]. Small colonies were counted directly and large colonies were counted using photographs. Also, some of these data have been previously presented by the author at congresses and articles [36, 38, 42].

2.1. Dupnisa Cave System

Dupnisa Cave System is located south of Sarpdere Village (Kırklareli) in Thrace, the European part of Turkey (**Figure 1**). The cave system lies in the forested Yıldız (Strandja) Mountains. Dupnisa Cave System is the second largest cave in Thrace region with a total length of 2720 m long. This cave system, which developed as a result of the disintegration of the Pliocene relief system in the upper part of the Yıldız Mountains with the Quaternary rivers, has the polycyclic development feature [31]. The cave system, which has four entrances, has developed horizontally and its formation process still continues. Dupnisa Cave System is regarded as a cave system because it is formed by two floors and three interconnected caves. These caves have different features. In this system, the active main gallery through which an underground stream flows is called Sulu Cave, while the totally fossilized ones above are called Kuru Cave and Kız Cave (**Figure 2**) [31].

Sulu Cave has the longest gallery of these caves with a length of 1977 m, and has only one entrance and one corridor. The height and width of this corridor are up to 40 and 15 m, respectively, and there is a very large hall of 125 m long, 80 m high and 35 m wide in this corridor. Kuru Cave, with a total length of 480 m, has two entrances and two corridors. In addition, a large hall was formed at the junction of these two corridors. Kuru Cave is connected with Sulu Cave by a narrow corridor. Kız Cave, 263 m in length, has one entrance and a small hall after the entrance. Kız Cave, which is covered with a thick fossil layer mixed with large blocks, gravel, and sand, is connected to Sulu Cave below at two points (**Figure 2**) [31].

Dupnisa Cave System is the first cave in the Thrace region that was opened to visitors in July 2003. This cave is also the first cave in Turkey to be opened to visitors with a program and gate construction according to the seasonal use of the cave by bats based on long-term-monitoring program [38]. Tourist circuits were constructed with the first 200 m of Sulu Cave and the first 230 m of Kuru Cave. However, Kız Cave is closed to visitors (**Figure 2**). The cave system has been visited by about 35,000 visitors each year after it was opened to visitors. The cave system has four entrances and two of these entrances, which are located on the tourist area, are closed to control human entry (**Figure 3**). Gates are constructed with a design of the horizontal angle iron bars that have 200-mm spacing between bars. The other entrances of the cave system, outside of the tourist area, where human entry is difficult, have been left to the natural state to minimize the negative effects of the two doors on the bats (**Figure 3**) [38].

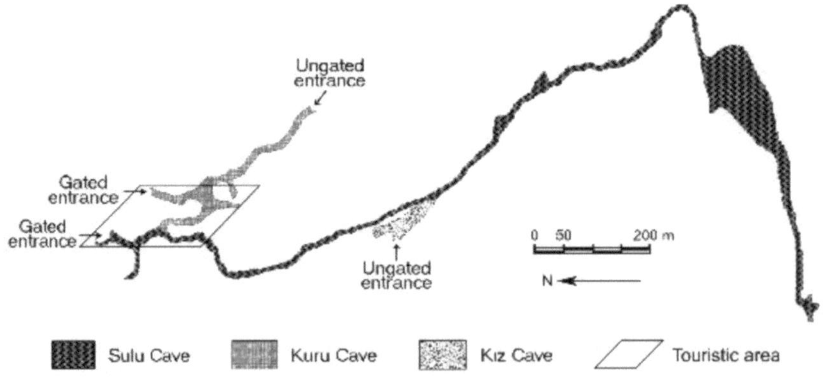

Figure 2. Dupnisa Cave System: the location of the three main caves, the areas open to tourists, and the gated and ungated entrances [38]. Adapted from Ref. [31].

Figure 3. The gated entrances of Sulu Cave (a) and Kuru Cave (b), used to control the entry of tourists. The ungated entrances of Kuru Cave (c) and Kız Cave (d), remained to minimize the negative effects of the gates on the bats.

2.2. Koyunbaba Cave

Koyunbaba Cave is located in the province of Kırklareli in Thrace of Turkey at 155 m altitude (**Figure 1**). Its length is 532 m, horizontally developed, and a fossil cave [31]. Depending on its development, the cave consists of different parts with various characteristics. There are seven stations used as roost by the bats in Koyunbaba Cave. These stations have different characteristics in terms of roost type, height, and microclimate (**Figure 4**).

Koyunbaba Cave developed on a prominent northwest-southeast direction fault line on the Eocene limestone. The fractured structure of the limestone resulted in the formation of many cavities and recess ledges in the cave. The cave initially developed in the northwest-southeast direction and takes the northeast-southwest direction from the middle section. At the intersection of these two sections, a large subsidence hall with a width of 50 × 60 m and a ceiling height of 30 m has developed (**Figure 4**). The formation of dripstone in Koyunbaba Cave, which has its base covered with a thick fossil layer consisting of soil, gravel, rubble block, and guano, is almost inexistent [31].

Koyunbaba Cave has three entrances, which are linked with different characteristics. The main entrance is the largest one. It is horizontally formed and its entry is very easy. Koyunbaba

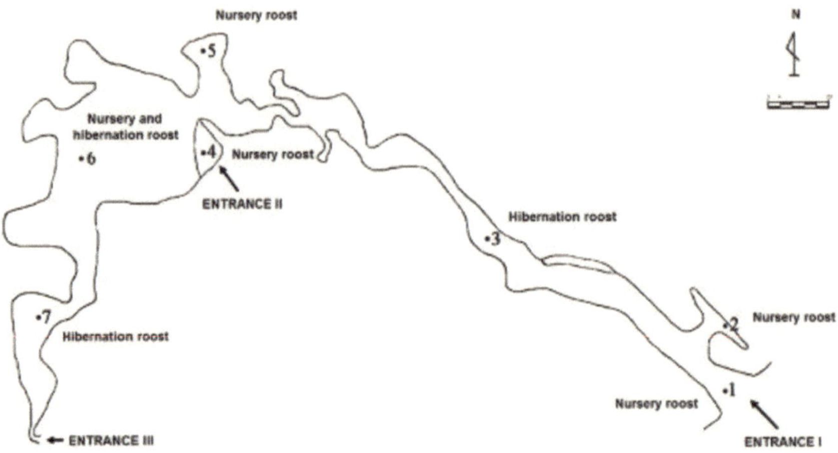

Figure 4. Koyunbaba Cave: the location of the seasonal roosts used by the bats, the three entrances of the cave. Adapted from Ref. [31].

Cave is closed to visitors, but this area of the cave is used by local people as an animal shelter during the summer. The second and third entrances of the cave are vertical. Because they are quite dangerous, these entrances cannot be used by local people (**Figure 5**). The cave is located in a relatively flat area consisting of farmland and pastures. This area is outside the Strandja Forest tree line. There are important water resources (Teke stream and Kayalı dam) around the cave.

Figure 5. The three entrances of Koyunbaba Cave: the main entrance (a), the second entrance (b), and the third entrance (c).

3. Results

3.1. Microclimate of Dunisa Cave System

Dupnisa Cave System, which has four entrances and three parts with different characteristics, provides seasonally different microclimatic conditions in the different parts (**Table 1**). Paksuz et al. [36] stated that the temperature varies between the different parts of the cave system, as well as by season. While the winter temperature ranges from −1 to 12°C in Sulu Cave, 8 to 14°C in Kuru Cave, and 10 to 14°C in Kız Cave, the summer temperature varies from 9 to 14°C in Sulu Cave, 13 to 22°C in Kuru Cave, and 13 to 18°C in Kız Cave. In winter, the mean temperature of Sulu Cave is significantly lower than from Kuru Cave and Kız Cave. During summer, the mean temperature in Kız Cave and Kuru Cave is significantly higher than in Sulu Cave. In Sulu Cave, the average temperature in winter is 7.4°C, in summer 10.7°C; in Kuru Cave in winter 11.4°C, in summer 16.3°C; in Kız Cave in winter 11.9°C, in summer 15.9°C. Humidity also varies depending on the caves and the seasons. The humidity in Dupnisa Cave System varies from 57 to 100%. The mean humidity in winter is similar and higher in Sulu Cave and Kız Cave, while it is lower in Kuru Cave. During summer, the humidity is similar in all three caves.

3.2. Bats of Dupnisa Cave System

Up to now, 17 species of bats have been recorded in Dupnisa Cave System, which is the most studied cave in Turkey in terms of bats. Five of these species belong to Rhinolophidae family (*Rhinolophus ferrumequinum, R. hipposideros, R. euryale, R. mehelyi*, and *R. blasii*) and 12 of them belong to Vespertilionidae family (*Myotis myotis, M. blythii, M. bechsteinii, M. emarginatus, M. nattereri, M. mystacinus, M. capaccinii, M. daubentonii, Miniopterus schreibersii, Barbastella barbastellus, Plecotus austriacus*, and *P. auritus*) [34–36, 38]. *M. alcathoe*, recorded in this study, has been newly registered for Dupnisa Cave System. The maximum number of bats recorded in Dupnisa Cave System is 54,600 in hibernation, while 11,000 in nursery. Dupnisa Cave System is used for hibernating by the majority (83%) of the total number of bats recorded, while it is used for breeding and nursing by the minority (17%). The three parts of the cave system were used by different species to varying degrees according to the season. Sulu Cave is used only for hibernating, while Kız Cave and Kuru Cave are used for both hibernating and nursery (**Figure 6**) [36, 38].

In Dupnisa Cave System, 99% of bat colonies are composed of six species, *M. schreibersii* (78%), *M. myotis/blythii* (8%), *R. euryale* (6%), *R. ferrumequinum* (4%), and *M. capaccinii* (3%).

Caves	Length (m)	Summer temp. (°C)	Winter temp. (°C)	Summer hum. (%)	Winter hum. (%)
Sulu Cave	1977	10.7	7.4	84.6	91.5
Kuru Cave	480	16.3	11.4	77.3	80.9
Kız Cave	263	15.9	11.9	80.7	87.6

Table 1. The length measurements and mean microclimate data of Dunisa Cave System.

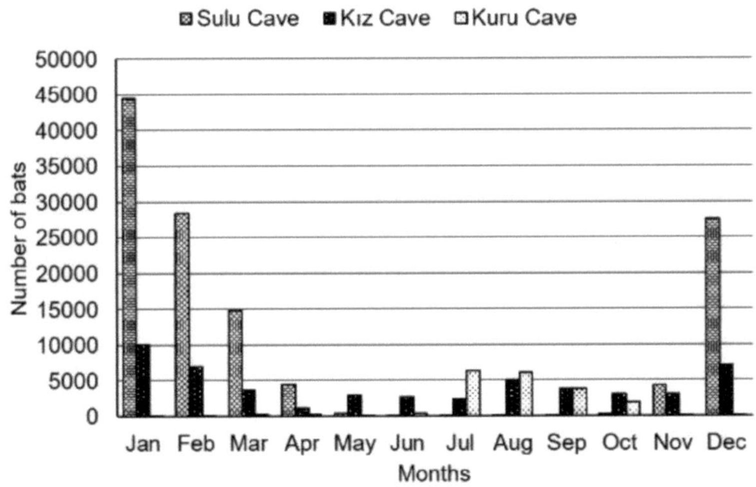

Figure 6. The maximum monthly number of bats recorded in a single survey in the three parts of Dupnisa Cave System. The data were collected from 2002 to 2010.

The biggest aggregations of the species recorded in a single survey in Dupnisa Cave System are *M. schreibersii* (45,600) in hibernation, *M. myotis/blythii* (4300) in hibernation, *R. euryale* (3600) in nursery, *R. ferrumequinum* (2200) in hibernation, *M. capaccinii* (1800) in hibernation, *R. mehelyi* (300) in nursery, *R. blasii* (200) in nursery, *M. emarginatus* (93) in hibernation, *R. hipposideros* (73) in hibernation, *M. daubentonii* (56) in hibernation, *M. mystacinus* (19) in hibernation, *M. bechsteinii* (six) in hibernation, *M. alcathoe* (five) in hibernation, *M. nattereri* (three) in hibernation, and *P. auritus* (one) in hibernation [36, 38].

While *Myotis* species mainly use Sulu Cave for hibernating, *Rhinolophus* species use both Sulu Cave and Kız Cave. On the other hand, *Myotis* species only use Kuru Cave for breeding and nursing, while *Rhinolophus* species use both Kuru Cave and Kız Cave. Although *M. schreibersii* is the most numerous species in Dupnisa Cave System, it does not use the cave system for breeding and nursing. However, unbred females and males of the species use Kuru Cave and Kız Cave only to spend the summer. Instead, *M. schreibersii* migrates to other suitable underground habitats in the area such as Koyunbaba Cave for breeding and nursing. In winter roosts, bat colonies are usually formed by one species. However, *M. myotis/M. blythii* and more rarely in smaller colonies *M. schreibersii/R. ferrumequinum* and *R. euryale/ R. mehelyi/R. blasii* can be seen together in the same roosts (**Figure 7**). On the other hand, summer colonies are of multi-species. Only *R. hipposideros* does not mix in with other species. In winter roosts, the two sexes are found together, whereas in summer roosts the two sexes are rarely found together.

The observation that the bats in the cave system always formed colonies at the same points depending on the seasons led to the belief that the rock characteristics, temperature, and humidity are the factors affecting roost selection. Paksuz et al. [36] emphasize that the main factor shaping the use of the cave system by the bats at various degrees is the microclimate. The temperature of Sulu Cave is low for breeding while the temperature of Kuru Cave is

Figure 7. The hibernation colonies of the bats in Dupnisa Cave System: a mixed colony of species *M. schreibersii* and *R. ferrumequinum* (a), a large colony of *M. schreibersii* (b).

high for wintering. Although the winter temperature is higher in Kız Cave than in Kuru Cave, Kız Cave is used for hibernating by the bats. The two entrances of Kuru Cave and its connection with Sulu Cave indicate that it may be more affected by temperature fluctuations and airflows, which may be the reason why the cave is preferred less than the other caves for hibernating. The humidity in winter is variable in all the three caves, while it is similar in summer. The specific roost requirements of bat species limit the use of many caves by bats. Thus, providing alternative conditions makes caves pretty appropriate roost for bats throughout the year. Therefore, Dupnisa Cave System, which offers alternative microclimatic conditions in the different parts, may be an important opportunity for many bats because the appropriate roosts will be more limited in the future, as global warming will become increasingly prevalent.

In Dupnisa Cave System, a conservation plan is applied for the protection of the bats and the cave system according to the seasonal use of the cave system by bats. This protection plan includes an appropriate visitor schedule and gate construction. The visitor schedule was arranged according to the seasonal use of Dupnisa Cave System by bats. The entrances of cave system where tourist circuits placed in were closed with horizontal angle iron gates to control the human disturbance. The other entrances of the cave system, outside of the tourist area, have been left to the natural state to minimize the negative effects of the two doors that can disturb the bats. In addition to these, some arrangements have been made for visitors and the use of lighting system. The protection of the caves and the bats will be possible only if the precautions that are taken and the suggestion that have been made are applied carefully [38].

Paksuz and Özkan [38] stated that the seasonal usage patterns of the parts in Dupnisa Cave System by the bats are completely preserved in periods of before and after tourist mobility (**Figure 8**). The authors also emphasized that there is no decrease in the total number of the bats in Dupnisa Cave System following the opening period tourist mobility. Moreover, they found a statistically significant increase after the tourist mobility only in Kız Cave, which is closed to tourism and ungated. This increase may indicate that the bats prefer to use the caves which are not visited by humans and tourist mobility. It seems as if Kız Cave, which is

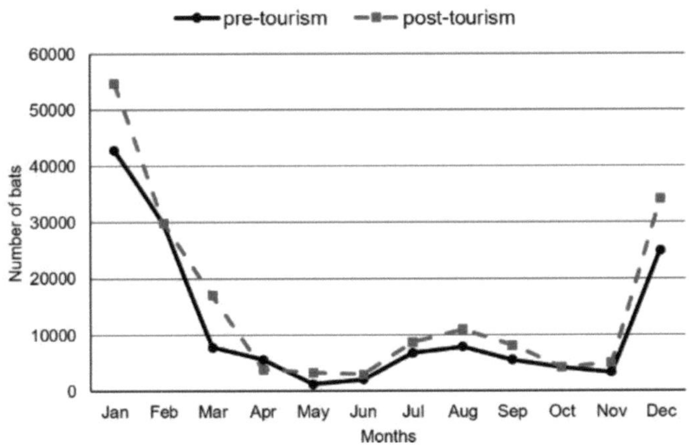

Figure 8. The monthly population sizes of the bats in the periods before and after tourist mobility in Dupnisa Cave System.

closed to tourism and ungated in Dupnisa Cave System, is a good opportunity to minimize the potential negative effects of the tourism activities in Sulu Cave and Kuru Cave on the bats. These results show that the protection program prepared for the protection of Dupnisa Cave System and bats is sustainable and must be applied meticulously.

3.3. Microclimate of Koyunbaba Cave

Koyunbaba Cave has different microclimatic conditions according to stations and seasons (**Table 2**) [37]. While the summer temperature in Koyunbaba Cave varies from 10 to 24°C, the winter temperature varies from 5 to 16°C. In summer, while the highest mean temperature is in station 4 (17.1°C), the lowest mean temperature is in station 7 (12.3°C). The highest (15.2°C) and lowest (5.2°C) mean temperatures in winter are in station 1. The humidity in Koyunbaba Cave varies from 58 to 98% depending on seasons and stations.

Stations	Roost location	Summer temp. (°C)	Winter temp. (°C)	Summer hum. (%)	Winter hum. (%)
1	Hall	17.0	9.3	66.7	68.6
2	Room	15.1	12.8	83.7	89.6
3	Corridor	12.6	8.1	75.9	86.0
4	Hall	17.1	9.4	68.9	76.2
5	Room	13.7	11.4	88.4	95.0
6	Hall	13.8	9.5	91.6	89.6
7	Corridor	12.3	7.9	75.7	80.4

Table 2. The mean microclimate data and location of the roosts in Koyunbaba Cave.

3.4. Bats of Koyunbaba Cave

In the studies conducted until now, 11 species of bats have been recorded in Koyunbaba Cave. Five of these species belong to Rhinolophidae family (*R. ferrumequinum, R. hipposideros, R. euryale, R. mehelyi,* and *R. blasii*) and six of them belong to Vespertilionidae family (*M. myotis, M. blythii, M. emarginatus, M. capaccinii, M. schreibersii,* and *P. austriacus*) [34, 37]. The maximum number of bats recorded in a single survey in Koyunbaba Cave is 29,500 in summer, while 600 in winter. Koyunbaba Cave is used by the majority (92%) of the total bat numbers for breeding and nursing, while it is used by the minority (8%) for hibernating (**Figure 9**) [37].

Seven species constitute 98% of the bat colonies in Koyunbaba Cave, *M. schreibersii* (55%), *M. myotis/blythii* (25%), *M. capaccinii* (7%), *R. mehelyi* (6%), *R. euryale* (3%), and *R. ferrumequinum* (2%). The largest aggregations of the species recorded in a single survey in Koyunbaba Cave are *M. schreibersii* (17,710), *M. myotis/blythii* (7840), *M. capaccinii* (2230), *R. mehelyi* (1850), *R. euryale* (859), *R. ferrumequinum* (788), *M. emarginatus* (356), *R. blasii* (176), *R. hipposideros* (10), and *P. austriacus* (one) [37].

The cave is used by species of *R. ferrumequinum, R. mehelyi,* and *R. hipposideros* for both breeding/nursing and hibernating throughout the year, while it is used by species of *R. blasii, R. euryale, M. myotis/blythii, M. capaccinii, M. emarginatus,* and *M. schreibersii* for breeding/nursing in the summer (**Figure 10**).

The roosts in Koyunbaba Cave are used by bats for various degrees and purposes according to species. The rooms and halls of Koyunbaba Cave are used as summer roost, while the corridors are used as winter roost. The selection of roost location in the cave by bats differed according to the species. *R. ferrumequinum, R. hipposideros, R. euryale, R. mehelyi,* and *M. emarginatus* use all roost types, while *R. blasii, M. myotis/blythii, M. capaccinii,* and *M. schreibersii* mainly use halls [37].

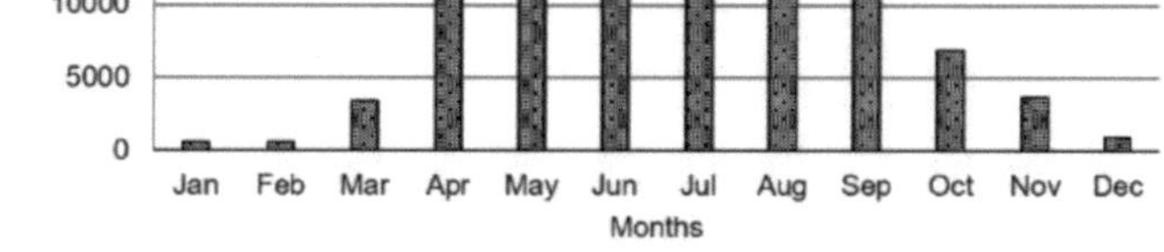

Figure 9. The maximum monthly number of bats recorded in a single survey in Koyunbaba Cave. The data were collected from 2007 to 2008.

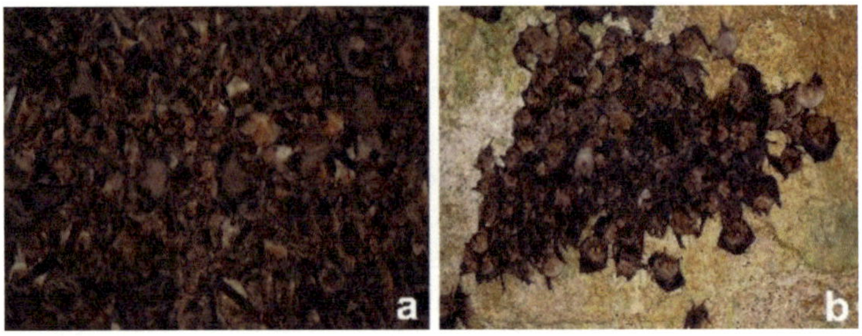

Figure 10. The breeding/nursing colonies of the bats in Koyunbaba Cave: a mixed colony of species *M. schreibersii, M. myotis, M. blythii,* and *M. capaccinii* (a), a nursery colony of Rhinolophidae family: grey(light)-colored bats are juveniles, brown(dark)-colored bats are adults (b).

These data show that Koyunbaba Cave is an important nursery roost for large populations of many bat species in the summer, as it has appropriate microclimatic conditions, large roosts, and entrances.

4. Conclusion

Turkish Thrace, located in the north-west of Turkey, served as a bridge between the Balkans, Anatolia, and Mediterranean. Thrace is one of the major biogeographic zones in Turkey and, due to its karst formation, there are many caves with different characteristics in the region. Due to these features, Thrace region, which has been an important glacial refuge for bats in the past, still provides a pretty opportunity to be hosted particularly for obligatory cave-dwelling bats. Turkish Thrace hosts large populations of many bat species [32–38].

Roost selection is essential for bats that spend more than half of their lives in roost [13]. Caves may serve as one of the most adequate roosting sites for bats because of the relatively stable microclimatic characteristics. Dupnisa Cave System and Koyunbaba Cave are the most important underground habitats for bat populations in southeastern Europe [34–38]. The hosting of large populations (56,600 hibernating bats and 11,000 breeding/nursing bats) of many bat species (18 species) is an indicator that Dupnisa Cave System is the most important shelter in the region. In addition to this, Koyunbaba Cave, which is used by a large breeding/nursing population (29,500) consisting of 11 bat species, is the most important summer roost in the region. Some bat species are listed as Near Threatened in the IUCN Red List, because of their declining populations in most of the European countries [43]. However, the populations of these species in Balkans and Turkey are stable [43], because Turkish Thrace provides many appropriate shelters to use by bats such as Dupnisa Cave System and Koyunbaba Cave.

Many species have very specific microclimatic requirements for roosts [18]. Microclimatic requirements of bats show seasonal variations according to their annual life cycle. Microclimatic conditions of roost and microclimatic requirements of species may contribute to patterns of association of bats [18, 23, 44]. Therefore, cave availability for bats is increased by providing alternative microclimatic conditions according to seasons. Most of the caves in

Turkish Thrace are appropriate shelters for winter colonies of bats and also for their nurseries. When there are alternative roosts, bats prefer the most appropriate ones to others. Therefore, Dupnisa Cave System is mainly used as a winter roost by bats, whereas Koyunbaba Cave is mainly used as a summer roost. Dupnisa Cave System and Koyunbaba Cave have different roost characteristics and microclimates and are used by different bat species for different purposes at different levels according to seasons. Therefore, they are alternative roosts to each other for bat population throughout the year. These caves, which complement each other in terms of seasonal use, are a chance for bats that have limited roosts. This makes these caves more important for the future of the bats in the region.

Growing urbanization across the world is resulting in negative impacts on bats and their key roosting habitats. Disturbance and destruction at caves is a widespread and major threat for cave-dwelling bats. Populations of some bat species are threatened globally due to human disturbance and roost lost caused by the increase in human population and land use. The protection of caves should be the most effective of bat conservation programs because a single cave can shelter thousands of bats from various species. Our findings related to these caves give some important clues for the population status and roost ecology of the bat species in Turkish Thrace. Also, it is very important for the implementation of any plans contributing to the protection of the roosts. The conservation of the caves, primarily Dupnisa Cave System and Koyunbaba Cave, whose importance for bats and other cave inhabitants was scientifically proven, is very important for the future of endangered species. Such preservation actions are made obligatory by international agreements including Turkey.

Acknowledgements

This chapter is produced from the following theses of the author. Master thesis (2004): Dupnisa Mağara Sistemi Yarasaları (Mammalia: Chiroptera). Doctoral thesis (2009): Koyunbaba Mağarası (Kırklareli-Türkiye) Yarasa Faunasının Mevsimsel Populasyon Değişimleri ve Tünek Seçimi.

Author details

Serbülent Paksuz

Address all correspondence to: serpaksuz@trakya.edu.tr

Faculty of Education, Trakya University, Edirne, Turkey

References

[1] Ford DC, Williams PW. Karst Geomorphology and Hydrology. London, UK: Unwin Hyman; 1989. 601 p. DOI: 10.1002/9781118684986

[2] Brown A, Pierson W, Brown K. Organic carbon resources and the payoff-risk relationship in cave ecosystems. In: Second International Conference on Ground Water Ecology; March 27-30, 1994; Atlanta, Georgia, USA: U.S. Environmental Protection Agency and the American Water Resources Association; 1994. pp. 67-76

[3] Culver DC, Pipan T. The Biology of Caves and Other Subterranean Habitats. Oxford: Oxford University Press; 2009. 256 p.

[4] Romero A. Cave Biology: Life in Darkness. Cambridge: Cambridge University Press; 2009. 306 p. DOI: 10.1017/CBO9780511596841

[5] Culver DC. Cave Life: Evolution and Ecology. Cambridge, Massachusetts: Harvard University Press; 1982. 189 p.

[6] Kunz TH, Racey PA, editors. Bat Biology and Conservation. Washington and London: Smithsonian Institution Press; 1998. 365 p.

[7] Pierson ED, Wackenhut MC, Altenbach JS, Bradley P, Call P, Genter DL, Harris CE, Keller BL, Lengus B, Lewis L, Luce B, Navo KW, Perkins JM, Smith S, Welch L. Species Conservation Assessment and Strategy for Townsend's Big-Eared Bat (*Corynorhinus townsendii townsendii* and *Corynorhinus townsendii pallescens*). Idaho, USA: Idaho Conservation Effort, Idaho Department of Fish and Game; 1999. 68 p.

[8] Hutson AM, Mickleburgh SP, Racey PA. Microchiropteran Bats: Global Status Survey and Conservation Action Plan. Switzerland and Cambridge, UK: IUCN/SSC Chiroptera Specialist Group, IUCN, Gland; 2001. 258 p.

[9] Altringham JD. Bats: Biology and Behaviour. Oxford, New York: Oxford University Press; 1996. 262 p.

[10] Simmons NB. Chiroptera. In: Wilson DE, Reeder DM, editors. Mammals Species of the World: A Taxonomic and Geographic Reference. 3rd ed. Vol. I. Baltimore: John Hopkins University Press; 2005. pp. 312-529

[11] Kunz TH, de Torrez EB, Bauer D, Lobova T, Fleming TH. Ecosystem services provided by bats. Annals of the New York Academy of Sciences. 2011;**1223**(1):1-38. DOI: 10.1111/j.1749-6632.2011.06004.x

[12] Schober W, Grimmberger E. The Bats of Europe and North America. New Jersey, USA: T.F.H. Publications; 1997. 239 p.

[13] Neuweiler G. The Biology of Bats. New York, NY: Oxford University Press; 2000. 310 p.

[14] Findley JS. Bats A Community Perspective. Cambridge: Cambridge University Press; 1993. 167 p.

[15] Nowak RM. Walker's Bats of the World. Baltimore: The Johns Hopkins University Press; 1994. 287 p.

[16] Anthony ELP, Kunz TH. Feeding strategies of the little brown bat, *Myotis lucifugus*, in southern New Hampshire. Ecology. 1977;**58**:775-786. DOI: 10.2307/1936213

[17] Whitaker JO Jr. Food of the big brown bat, *Eptesicus fuscus*, from maternity colonies in Indiana and Illinois. American Midland Naturalist. 1995;**134**:346-360. DOI: 10.2307/2426304

[18] Kunz TH. Roosting ecology of bats. In: Kunz TH, editor. Ecology of Bats. New York, NY: Plenum Press; 1982. pp. 1-56

[19] Chaverri G, Quiros OE, Gamba-Rios M, Kunz TH. Ecological correlates of roost fidelity in the tent-making bat *Artibeus watsoni*. Ethology. 2007;**113**:598-605. DOI: 10.1111/j.1439-0310.2007.01365.x

[20] Humphrey SR. Nursery roosts and community diversity of nearctic bats. Journal of Mammology. 1975;**56**:321-346.DOI: 10.2307/1379364

[21] Tuttle MD. Population ecology of the gray bat (*Myotis grisescens*): Philopatry, timing and patterns of movement, weight loss during migration, and seasonal adaptive strategies. Occasional Papers of the Museum of Natural History, University of Kansas. 1976;**54**:1-38

[22] Fitzgerald JP, Meaney CA, Armstrong DM. Mammals of Colorado. Denver: Denver Museum of Natural History and University Press of Colorado; 1994. 467 p.

[23] Baudinette RV, Churchill SK, Christian KA, Nelson JE, Hudson PJ. Energy, water balance and the roost microenvironment in three Australian cave-dwelling bats (Microchiroptera). Journal of Comparative Physiology B. 2000;**170**:439-446. DOI: 10.1007/s003600000121

[24] Smirnov DG, Vekhnik VP, Kurmaeva NM, Shepelev AA, Il'in VY. The spatial structure of the community of Chiroptera (Chiroptera: Vespertilionidae) hibernating in artificial caves of the Samarskaya Luka. Izvestiia Akademii nauk. Seriia Biologicheskaia. 2008;**2**:243-252

[25] Siivonen Y, Wermundsen T. Characteristics of winter roosts of bat species in southern Finland. Mammalia. 2008;**72**:50-56. DOI: 10.1515/MAMM.2008.003

[26] Jones G, Jacobs D, Kunz TH, Wilig MR, Racey PA. Carpe Noctem: The importance of bats as bioindicators. Endangered Species Research. 2009;**8**:3-115. DOI: 10.3354/esr00182

[27] Casale A, Vigna Taglianti A. Caraboid beetles (excl. Cicindelidae) of Anatolia, and their biogeographical significance (Coleoptera, Caraboidea). Biogeographia. 1999;**20**:277-406. DOI: https://doi.org/10.21426/B620110002

[28] Nazik L, Törk K, Tuncer K, Özel E, İnan H, Savaş F. Türkiye Mağaraları. In: Ulusal Mağara Günleri Sempozyumu Bildiri Kitabı; 24-26 Haziran 2005; Beyşehir, Konya: Türkiye Tabiatını Koruma Derneği; 2005. pp. 31-46

[29] Kültür ve Turizm Bakanlığı TC. Mağara Turizmi [Internet]. 2017. Available from: http://www.ktbyatirimisletmeler.gov.tr/TR,10336/turizme-acik-magaralar.html [Accessed: January 8, 2017]

[30] Ekmekçi M. Karst in Turkish Thrace: Compatibility between geological history and Karst type. Turkish Journal of Earth Sciences. 2005;**14**:73-90

[31] Nazik L, Törk K, Özel E, Mengi H, Aksoy B, Acar C. Kuzey ve Kuzeydoğu Trakya'nın (Kırklareli-Tekirdağ) Doğal Mağaraları. Dosya No: 43584. Ankara: Maden Tetkik ve Arama Genel Müdürlüğü, Jeoloji Etütleri Dairesi Başkanlığı; 1998

[32] Bilgin İR. The summer populations of cave-dwelling bat species of Çatalca-Kocaeli region and environmental factors that influence their distribution [MS thesis]. Istanbul: Boğaziçi University; 2000

[33] Furman A, Özgül A. Distribution of cave-dwelling bats and conservation status of underground habitats in the Istanbul area. Ecological Research. 2002;**17**:69-77. DOI: 10.1046/j.1440-1703.2002.00468.x

[34] Furman A, Özgül A. The distribution of cave-dwelling bats and conservation status of underground habitats in Northwestern Turkey. Biological Conservation. 2004;**120**:243-248. DOI: 10.1016/j.biocon.2004.02.019

[35] Paksuz S. Dupnisa Mağara Sistemi Yarasaları (Mammalia: Chiroptera) [MS thesis]. Edirne: Trakya University; 2004

[36] Paksuz S, Özkan B, Postawa T. Seasonal changes of cave-dwelling bat fauna, and their relationship with microclimate in Dupnisa Cave System (Turkish Thrace). Acta Zoologica Cracoviensia. 2007;**50A**(1-2):57-66. DOI: 10.3409/000000007783995435

[37] Paksuz S. Koyunbaba Mağarası (Kırklareli-Türkiye) Yarasa Faunasının Mevsimsel Populasyon Değişimleri ve Tünek Seçimi [PhD thesis]. Edirne: Trakya University; 2009

[38] Paksuz S, Özkan B. The protection of the bat community in the Dupnisa Cave System, Turkey, following opening for tourism. Oryx. 2012;**46**(1):130-136. DOI: 10.1017/S00306 05310001493

[39] Çoraman E, Çelik YE. Identification and Protection of Important Bat Caves in Turkey [Internet]. 2012. Available from: http://www.conservationleadershipprogramme.org/media/2014/12/0534710_Turkey_FinalReport_Identification-of-importnat-bat-caves_15thNov2013.pdf [Accessed: August 12, 2016]

[40] Çoraman E, Çelik YE, Irmak S. Yarasalar & Mağara Ekosistemleri [Internet]. 2012. Available from: http://www.milliparklar.gov.tr/magara/m_yarasalar.pdf [Accessed: June 10, 2016]

[41] Dietz C, Von Helversen O. Identification Key to the Bats of Europe – Electronical Publication, Version 1.0 [Internet]. 2004. Available from: www.unituebingen.de/tierphys/Kontakt/mitarbeiter_seiten/dietz.htm [Accessed: December 15, 2004]

[42] Paksuz S. Microclimatic preferences of bat species in Koyunbaba Cave (Kırklareli-Turkey). In: Proceedings of International Scientific Conference-UNITECH; November 18-19, 2016; Gabrovo: Technical University of Gabrovo Bulgaria; 2016. pp. 429-434

[43] Hutson AM, Aulagnier S, Benda P, Karatas A, Palmeirim J, Paunović M. *Miniopterus schreibersii*. IUCN Red List of Threatened Species. Version 2009.1 [Internet]. 2009. Available from: http://www.iucnredlist.org [Accessed: December 2, 2009]

[44] Borda D, Borda C, Tămas T. Bats, climate, and air microorganisms in a Romanian cave. Mammalia. 2004;**68**(4):337-343. DOI: 10.1515/mamm.2004.033

Section 3

Geophysical Application for Cave Detection and Geoecological Assessments of Water Chemistry in Karst Aquifers

Chapter 5

Investigation of Water Quality in the Agricultural Area of Lithuanian Karst Region

Aurelija Rudzianskaitė

Additional information is available at the end of the chapter

http://dx.doi.org/10.5772/intechopen.68536

Abstract

In North Lithuania, karst processes are developing in the formations of the Upper Devonian represented by fissured gypsum-dolomite. The result of these processes on surface is the formation of sinkhole. The precipitation infiltrates into the karst aquifer not only through Quaternary sediments but also through karst sinkholes. The groundwater is extremely vulnerable to pollution in karst region. Agricultural practise is prevailing in this region. The objective of chapter is present of the results from various investigations on drainage, rivers, sinkholes and ground water quality in the agricultural area of Lithuanian karst region. This study was prepared on the basis of scientific publication analysis and the generalization methods were used. Based on this result, chemical composition of research water is mostly depended on the type of soil, meteorological conditions and land used. Higher amount of nitrogen and phosphorus are leached out in winter and spring periods.

Keywords: agriculture area, nitrogen compounds, total phosphorus, sinkhole, water quality

1. Introduction

The karst phenomena occur in different countries (Central America, North and South China, Slovenia, Estonia, Latvia etc.). The water in a karst aquifer is a major water resource in many regions of some countries [1].

The karst region is located in North Lithuania. Karst processes occur in carbonate rocks (gypsum-dolomite) forming different surface and underground karst features such as sinkholes, caves and caverns. Gypsum is soluble and hence its layers can dissolve and the protective cover strata of karst rock become unstable and may subside. In region, gypsum occurs in contact with water and this causes subsidence problems. In agriculture land, these phenomena are inconvenient, but in urban area, they constitute a geological hazard that can seriously affect development and human safety [2].

The rate of chemical denudation of soluble rocks is one of the main factors determining the intensity of karst process. Gypsum denudation is mainly predetermined by water balance, which depends on meteorological conditions (precipitation and evaporation). The highest intensity of gypsum denudation is during the spring flood and the lowest in the dry period. Since 1978, the intensity of karst denudation has been increased by 30% [2, 3].

The main human activities influencing karst development are groundwater extraction and agriculture [4, 5].

Karst region is densely populated. The two districts (Biržai and Pasvalys) with more than 53,000 of inhabitants [6] are located in it. Soils are fertile; agriculture has been a traditional activity on these soils. Precipitation water dissolves not only natural salts from soils but also nutrient surplus from agriculture areas where unbalanced fertilisation rate was applied. Problems are caused by filtration of the water which is rich in salts and which pollutes groundwater. The major amount of precipitation infiltrates into the karst aquifer through Quaternary sediments, however, infiltration takes place through karst sinkholes as well [3, 7]. Karst groundwater becomes polluted frequently and in shorter time periods than water in non-karstic aquifers [1].

The purpose of this chapter is to present a short review of the research finding on water (drainage, stream, groundwater and sinkhole) quality in agricultural area of North Lithuanian karst region.

2. Geology of the North Lithuanian karst region

The total area of Lithuania is 65,300 km^2 and is situated along the south-eastern shore of the Baltic Sea [8]. The karst region is located in North Lithuania (mainly in Biržai and Pasvalys administrative districts) and borders with the karst territories of the neighbouring South Latvia (**Figure 1**).

The karst processes highly active are related to Upper Devonian gypsum and dolomites that occur beneath the Quaternary cover (**Table 1**). Mainly of Quaternary deposits consist of loam and sandy loam, sand, gravely sand, gravel clay and silt; in some places, karstified rocks are overlain by the Pamūšis Formation (dolomitic marl, clay and other) and the Istras Formation (dolomite) of Upper Devonian [4, 7, 9].

Karst in the North Lithuania is matured not only at the land surface (sinkholes, karst shafts and dolines) but also in the subsurface (cavities and caves). In the region of active gypsum, karst (the area of 400 km^2) is counted more than 8500 different size and shape sinkholes [4, 7, 9].

The thickness of the cover of the karst rocks in the area varies from parts of the first metre up to 70 m and has a common tendency to increase southwards and especially westwards.

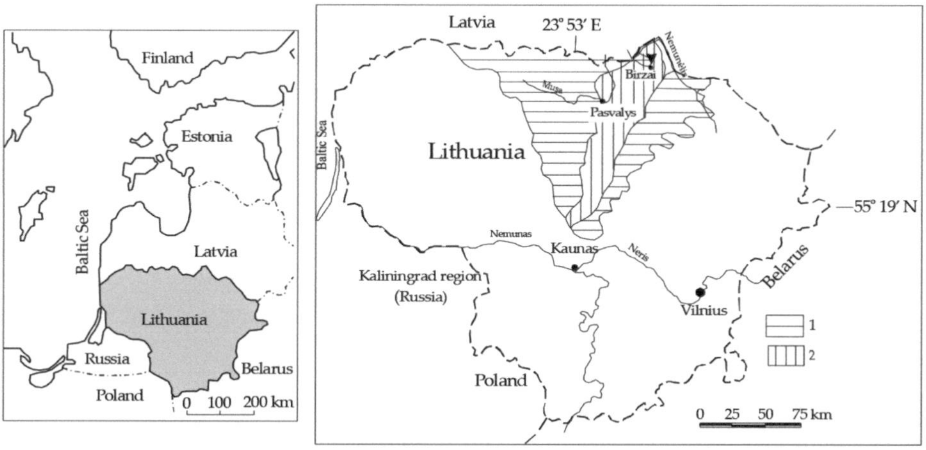

Figure 1. Location of Lithuanian karst area: (1) Pre-quaternary carbonates; (2) area of intensive karst in North Lithuania (according to literature [2, 3]).

System	Series	Formation	Index	Thickness, m	Description of sediments
Quaternary			Q	0–20	Moraine loam
Devonian	Upper	Pamusis	D_3pm	15	Clayey marl
		Istras	D_3ys	3–9	Dolomite fissured
		Tatula	D_3tt_n	11–15	Gypsum interlayers with marl and dolomite
			D_3tt_k	3–7	Marl clayey
			D_3tt_p	13–24	Gypsum interlayers with marl and dolomite
		Pliavinia	D_3kp	6–12	Dolomite fissured
			D_3ss	13–18	Dolomite clayey
			D_3j	2.0–9.9	Dolomite, marl
		Sventoji	D_3sv	90	Sand interlayer with siltstone and sandstone
	Middle	Upninkai	D_2up	70–110	Sand clayey and clay
		Narva	D_2nr	~100	Clayey-carbonate deposits

Table 1. Stratigraphy of the Middle-Upper Devonian sediments in the North Lithuania [4].

Sinkholes are located in the areas with the cover thickness less than 25 m. In the areas where the cover thickness is up to 5 m, new sinkholes are forming intensively [9, 10].

3. Hydrogeology and hydrology of the North Lithuanian karst region

The North Lithuania karst region is located in the eastern part of the Baltic Artesian Basin. The active water change zone is up to 270 m thick and includes aquifers in the Quaternary and in the Upper Devonian (the Istras-Tatula, Kupiskis-Suosa and Sventoji-Upninkai) formations. This series of aquifers is underlain by the 60–100 m thick regional aquitard of the Narva Formation. The main source for recharge of the Istras-Tatula and Kupiskis-Suosa aquifers is shallow groundwater and surface water. All aquifers are being exploited to a different extent for drinking and domestic purposes and for industrial needs in this region [4, 11]. The infiltration rate of precipitation influences variations in level and chemical composition of groundwater. Ground collapse affects the Quaternary cover and permits ready recharge of surface water into the Upper Devonian aquifers. The intensive karst zone is referred to the areas with intensive water circulation in open gypsum systems [4, 11, 12]. Water of the aeration zone mostly infiltrates through vertical fissures until it reaches either horizontal channels or impermeable soil layer. A large amount of precipitation is collected in the bottom of the sinkholes, which are open holes or filled with permeable deposits [7].

In the karst region, most of the sinkholes are dry. Only during spring floods, they collect and temporarily retain atmospheric water or groundwater. In the course of time, the newly occurring sinkhole becomes shallow as a result of sedimentation processes and transforms into small bogs. As a result of a very good hydraulic link between the surface and groundwater, part of sediments is eliminated with the ground run-off [3]. Karst lakes and dry sinkholes make natural drainage systems through which precipitation, snow melting and overland flow access gypsum strata of the Lower Devonian [2, 12].

Karst rivers are recharged mainly through the precipitation. A low part of run-off is formed by the karst groundwater. Many rivers in the region cut upper part of the karst rocks and aquifers [4]. The groundwater discharge into the karst rivers of the region makes up from 25 to 40%, whereas in non-karst rivers it comprises only 8–16% of the annual run-off. The run-off of karst rivers in drought periods always exceeds the run-off of non-karst rivers [11].

4. Meteorological condition

The climate of the karst region as everywhere in Lithuania is intermediate between the maritime and continental climates. The annual infiltration of precipitation exceeds evaporation [8].

Air temperature observations in the active karst zone were started only in 1924 [12].

According to the data of Biržai Meteorological Station, the long-term annual averages of the main climatic characteristics are as follows: air temperature 6.6°C; precipitation 661 mm. The mean temperature for July, the warmest month of the year, is 18°C, and for January, the coldest month, it is 3.6°C below zero. The rainfall during a warm season, from April to October, makes up 80% of precipitation. Long-term mean monthly amount of precipitation is varying from 36 to 81 mm.

Snow appears at the end of November and melts at the end of March. Thickness of snow cover reaches 6–44 cm. The mean depth of freezing is varying 17–99 cm.

During the last decades of the twentieth century, winter season temperature particularly near ground surface has increased significantly [12]. A distinct increase in air temperature has been observed since 1960. In karst region, the average temperature of January and February increased in the period 1961–1990 compared with the period 1931–1960 by 0.5°C [13].

The researchers [12] state that mean temperature in summer decreased until 1980 and then started to increase. Significant increases in annual, winter and summer temperatures were observed in 1994–2004. Increases in air temperature, during both cold and warm seasons, influenced other hydrometeorological characteristics and the solubility of gypsum.

5. Water quality

5.1. Agricultural drainage

Agriculture has been a traditional activity on soils of North Lithuania karst zone, whereas these soil are comparatively fertile. Agricultural areas in mainly administrative district Biržai make up 69% and in Pasvalys it is 79.9%, whereas in Lithuania it makes average 61.3% [14].

Lithuania is in the zone of surplus humidity, therefore, agricultural areas need soil drainage. The total drained land area occupies 47% of the Lithuanian land area and 86% of the agricultural land area, of which 87% is tile-drainage [8]. The leaching of nutrients from agricultural fields is one of surface and groundwater pollution sources. The most chemical matter contained in the soil is water-soluble; therefore, hydrological regime changes in the ecosystem are closely related to changes in nutritive leaching. By means of drain, much nitrate (NO_3) and other elements contaminating the stream (drainage receivers) water are leached out. The study results [15, 16] in moraine sandy loam and glacial-lacustrine clay soil show that the drainage water influenced on water quality of streams via drained agricultural areas. As a matter of fact, the nitrogen amount leached via drainage, which has impacts on the concentration quantity in stream water (**Figure 2**). NO_3 amount leached out by the drain water (determination coefficient 0.28–0.89), influence concentration fluctuations of the same compound in the stream. In view to phosphorus amount in the stream water, it was the result of other sources and factors. It was established [15] that stream water mineralization was greater, and its chemical composition depended on soil texture, fertilizers application and intensive land use in the drained areas. Mostly higher nitrate concentration was under the conditions of run-off formation after the dry period of the year; higher total phosphorus concentrations occurred when higher precipitation amount fall, particularly in summer.

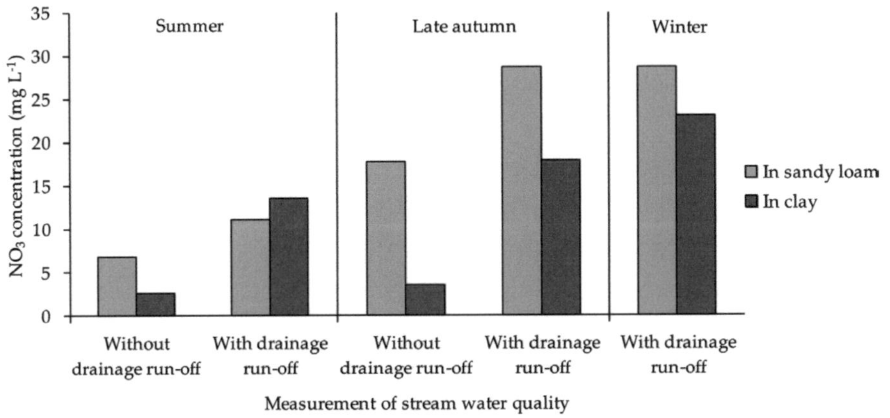

Figure 2. Dependence of nitrate concentration in the stream water on the drainage function.

It is known that the water quality depended on cultivated crops in this area. The highest concentration of nitrate nitrogen (NO_3–N) was found when growing potatoes and lower one was found when growing perennial grasses (**Figure 3**). However, concentration of total phosphorus (TP) increases in drainage water when growing perennial grasses [17, 18].

The studies [19] were carried out in spring crop fields (barley with perennial grass under crop) in sandy light loam and clay concluded that leaching of nitrate nitrogen and chlorine is mostly determined by fertilization rates; the effect of fertilization on leaching of potassium is less significant.

For Lithuanian conditions, the potential impact of climate change is associated with alterations in the precipitation pattern turning from snowfall to rainfall [20]. As a result, tile-drainage run-off volume tends to increase in the winter months.

The study [21] of clay and clay loam in the Lithuanian karst zone demonstrated that the average air temperature and precipitation from the period 1988 to 1999 exceeded the corresponding data from the period 1976 to 1987 by 1.3°C and 34 mm, respectively. A significant increase in the air temperature was observed in January and February (4.8°C). For this reason, a 3.3

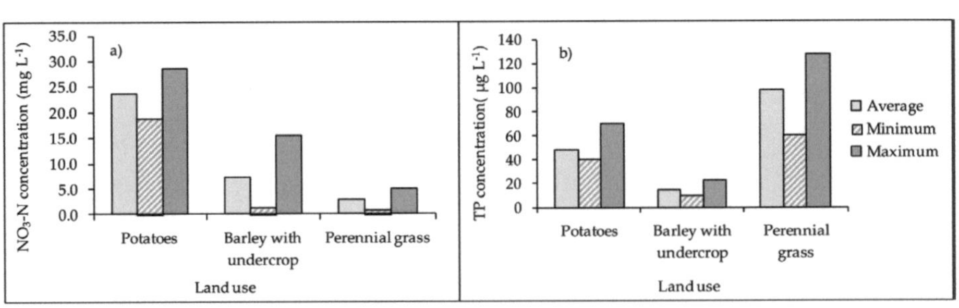

Figure 3. Nitrate nitrogen concentration (a) and total phosphorus concentration (b) in sandy soil water.

times higher drainage run-off volume was observed in the winter (January–February) and 1.6 times lower drainage run-off volume was identified in the spring (March–May) during the period 1988–1999 compared with the period 1976–1987. Concerning the changes of drainage hydrological regime in the cold period, the new environmental problems related to the increase of the leaching of nutrients were found.

Higher amount of nitrogen and phosphorus are leached out in winter and spring periods [18]. The peak concentrations of nitrate nitrogen occurred in winter and spring (due to the higher run-off) and at the onset of run-off after a dry period (due to the leaching of nitrogen that was not used during the dry period) [22].

5.2. River

The convection is mainly affected by transport processes of nutrients in karst region. In the first, nutrients with precipitation infiltrate via the aeration zone into the shallow aquifer and ending with surface or subsurface flow into the rivers [23].

Run-off and water quality of rivers are determined by a number of factors such as physical geographical conditions of the catchment climatic factors, human economic activity [24–26]. Inertia is a characteristic to the inflow of biogenic materials into the rivers. It is determined by forestless of the catchment, chemical properties of the soils and humus content [27]. In the last decade, the Mūša River catchment (karst region river) has received about 87% of the total nitrogen from arable land, waste water treatment, households and urban territories—10% and from wooded territories and pastures—3% [28]. Although agricultural production, use of mineral fertilizers and pesticides has decreased, concentration of nitrogen compounds in the rivers of agricultural influence has increased 2.5–3 times due to incorrect handling of manure, changes in the structure of agricultural lands and agrotechnics [29]. The researchers [30] stated that the stream pollution associated with agricultural activity, significantly water purified already at a 1.5 km stretch when stream flowing through forest-covered area. The average nitrogen concentration decreases about 3.0 times a day during the vegetation period and about 2.6 times a day during cold season of the year.

The highest total nitrogen amounts (about 36%) got to the rivers during the winter period (January–December), when there is no vegetation processes. In spring (March–April), in the period of snow melting, one third or 31% was scoured because of a higher run-off to the researched subcatchment and in the summer (May–August) and autumn seasons (September–November), 16 and 17% of all nitrogen falling to the subcatchment, respectively [31].

North Lithuania karst zone has specific geomorphological conditions. Here, the sinkholes that formed in the course of karst processes create favourable conditions for the interaction of surface and ground water. For example, direct zone of ground water nutrition is in the watershed of the Tatula and Apaščia Rivers. There exist more internal run-off zones of this kind, where most of the surface run-off gets to the horizons of soluble rocks [32]. Karst phenomena have the strongest impact on chemical composition of water in the middle reaches of the Tatula River and the lower reaches of the Lėvuo River [33].

Analysis of investigation results of small rivers of agricultural influence and big rivers of state monitored agricultural influence has established that bigger rivers have more intensive agitation processes in the flow and receive run-off from greater areas, while quality of small streams is under immediate impact of local factors and distinguished impact of particular factors in the catchments [34].

The peculiarities of hydrological regime and pollution were observed in an analysis of data from 1996 to 2005 in the rivers (Lėvuo, Tatula and Nemunėlis) of the karst region. Comparative analysis of water quality in rivers and stream (the Bėrė, tributary of the Mūša and the G-1, tributary of the Apaščia) is completed [35]. Only agricultural activity is developed in the catchments of these streams, there are no big settlements or concentrated pollution sources (**Table 2**).

River/stream	Catchment area (km)	Forest (%)	Arable land (%)	Pasture (%)
Levuo*	1629	29	56	11
Tatula	453	14	73	10
Nemunelis*	1892	15	51	30.6
Bere**	2.02	24	66	8
G-1**	1.63	14	46	35

*Catchment area in Lithuania country.
**Catchment area to the water quality study site.

Table 2. Description of study catchment of rivers and stream.

In the period of investigations, concentration of nitrate nitrogen is up to 2.1 times smaller in the rivers than it is in the stream (**Table 3**). In streams of agricultural influence (the Bėrė and the G-1), cases of good ecological condition (NO_3-N = 1.3–2.3 mg L^{-1} [36]) make 33–38% of total number of measurements. Concentration of ammonium nitrogen (NH4-N) was similar and low in all the investigated rivers. During the period of investigations, the percentage of cases of good ecological condition (NH_4-N = 0.10–0.20 mg L^{-1}) varied from 48 to 66.

		Lėvuo	Tatula	Nemunėlis	Bėrė	G-1
NO_3-N	Minimum	0.04	0.04	0.02	0.12	0.26
	Maximum	11.0	11.9	13.0	63.0	10.5
	Average	3.08	3.25	2.82	6.01	3.67
NH_4-N	Minimum	0.004	0.007	0.007	0.006	0.002
	Maximum	4.9	11.0	4.6	4.12	2.93
	Average	0.518	0.485	0.364	0.401	0.306
TP	Minimum	0.025	0.01	0.024	0.016	0.004-
	Maximum	8.25	1.16	5.35	2.1	0.19
	Average	1.139	0.144	0.232	0.122	0.045

Table 3. Nitrate nitrogen (NO_3-N), ammonium nitrogen (NH_4-N) and total phosphorus (TP) concentrations in the river and stream water (mg L^{-1}).

Higher concentrations of total phosphorus are established in the Lėvuo and the Nemunėlis Rivers. Catchments of these rivers have more built-up territories, which may discharge various wastewaters. In the Bėrė stream, higher concentrations of total phosphorus might be determined by phosphorus content in the soil and rich fertilization.

Correlative analysis has established stronger correlation between concentrations of nitrate nitrogen in rivers and concentrations of ammonium nitrogen in streams, which assumes that change of these elements has common regularities. Concentrations of total phosphorus have very weak correlations, thus the reasons of their occurrence in water of the rivers should be different.

5.3. Sinkholes

Sinkhole water quality may worsen due to accumulation of organic materials (increasing peat content) or natural changes of environment (abrasion of slopes, overgrowth, etc.), while anthropogenic impact is minimal [37]. Sinkholes create favourable conditions for interaction of surface and groundwater. Identification of possible sources of sinkholes pollution is very important in the solution of groundwater safety problems.

The studies of water quality in sinkholes were carried out in the active karst zone in 2008–2012 (**Figure 4**) [38].

The research sinkholes were of different ages and have differently overgrown slopes (**Table 4**).

In most cases, nitrate nitrogen (0–0.97 mg L^{-1}) and total phosphorus (0.017–0.042 mg L^{-1}) concentrations in water of sinkholes were similar to this concentration in precipitation.

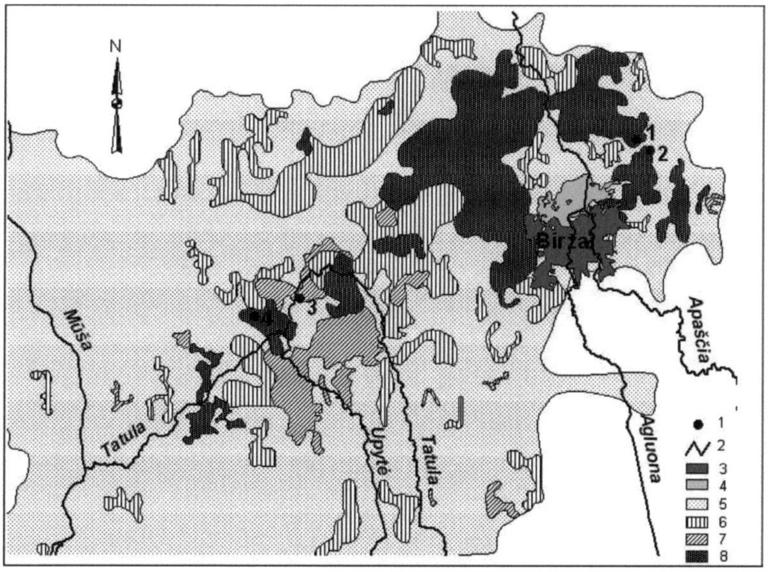

Figure 4. Scheme of location of studied sinkholes: (1) studied sinkholes; (2) river; (3) Biržai town; (4) Širvena Lake; (5–8) land group (5—first, 6—second, 7—third, 8—fourth).

Sinkhole No	Description
1	The sinkhole (area is 0.105 ha) was old age (emerged 100–1000 years ago [39]) and peat-filled, covered with marsh vegetation, it was usually dry in the summer. The sinkhole was surrounded by meadows, at 25 m distance from the stream G-1
2	The sinkhole (few merged sinkholes) was old and peat-filled, overgrown with trees and shrubs, it is dry in the summer. The trees and shrubs covered area was 0.572 ha
3	The sinkhole (area is 0.021 ha) was a new one (emerged in the last 20 years). Water regularly filled it about 2 m from the ground surface. Slopes were mineral ground, mostly overgrown with grass vegetation, but fast growing shrubs, abrasion going on the slopes. The sinkhole was surrounded by meadows
4	The sinkhole (area is 0.057 ha) was young (emerged 20–100 year ago), water filled about 0.5 m above the ground surface, the coast was covered with wetland vegetation. In this environment, there were many dry and old sinkholes covered with grasses

Table 4. Description of the research sinkhole.

Water quality of sinkholes depend on age sinkholes and slopes ground. The higher amounts of nitrate nitrogen (0–32.2 mg L^{-1}), total phosphorus (0.36–1.32 mg L^{-1}), sulphate (2.7–172.0 mg L^{-1}) and the amount of organic matter according to biochemical oxygen demand (1.91–76.0 mg O_2 L^{-1}) were in water of peat-filled sinkholes than in water of sinkholes with mineral ground slope (young sinkholes) (**Table 5**). Water of sinkhole that is overgrown with trees and shrubs has the highest nitrate nitrogen (up to 32.2 mg L^{-1}) and sulphate (up to 172 mg L^{-1}) concentration.

		Sinkhole No				Precipitation [22]
		1	2	3	4	
NO_3-N	Minimum	0	0	0	0	0.04
	Maximum	0.97	32.2	0.189	0.059	2.21
	Average	0.10	11.39	0.02	0.01	0.73
TP	Minimum	0.042	0.036	0.017	0.013	0.021
	Maximum	1.32	1.03	0.132	0.118	0.301
	Average	0.42	0.35	0.05	0.05	0.111
SO_4	Minimum	2.7	2.0	1.4	1.2	-
	Maximum	-50.0	172.0	42.0	31.0	
	Average	15.4	82.7	16.4	9.4	
BOD	Minimum	2.13	1.91	1.56	1.41	-
	Maximum	32.0	16.90	14.20	11.0	
	Average	12.61	8.45	4.0	1.41	

SO_4 – sulphate.
BOD – biochemical oxygen demand.

Table 5. Sinkholes and precipitation water quality (mg L^{-1}).

Analysing [40] the regime of water in four peat-covered sinkholes and groundwater of their vicinity, it was determined that during the periods of snow thaw and heavy rains the sinkholes receive subsurface and ground water from the vicinity, while during the dry period water accumulated in peat media of the sinkholes flows into the vicinity. The water in sinkholes gets abated and the sinkholes were draining groundwater intensively during dry period. The investigation [41] in the dry period shows that water level of peat-filled sinkholes was higher and nitrate nitrogen, chlorine and sulphate concentration were smaller. In this period, the sinkhole water percolating into a stream, it refills and at the same time attenuates stream water and makes its quality better, although a larger concentration of ammonium, nitrate and total phosphorus makes it worse.

5.4. Groundwater

Problems are caused by filtration of the water rich in salts, which pollute groundwater. Polluted drinking groundwater is dangerous for people's health. The salt-saturated water becomes more aggressive and intensifies karst process of the rocks lying below, which definitely causes a risk to people's property or event life [42].

The rural dwellers (53% of all rural population in Birzai and 70% in Pasvalys districts) are mainly using the ground water from individual dug wells. There are over 19,000 dug wells in Biržai district. They can be characterized as wells in which the admissible water pollution rates are exceeded two or more times. In 2004, microbiological analyses were carried out in 77 town and 278 village wells. The analyses showed that drinking water did not correspond to the hygienic standard in 28.5% of town and 20.5% of village wells. Town wells had increased chemical pollution and village wells had increased microbiological pollution of water [43]. The nitrate concentration in wells in the moraine sandy loam was higher than in the glacial lacustrine clay and it exceeded the highest allowed value (HAC 50 mg L^{-1}) from 1 to 33 times (**Figure 5**). The highest quantity of organic substances (according to permanganate oxidation) were found in the wells of glacial lacustrine clay and shallow wells (moraine sandy loam) [44]. The greatest amounts of organic substance were determined when the level of water fluctuated near to the earth's surface (could get together with a surface flow) and near to the ground of well (the amount of organic substances in the water has increased through ground deposits) [45]. The highest pollution of well water with organic matters is observed in summer [46].

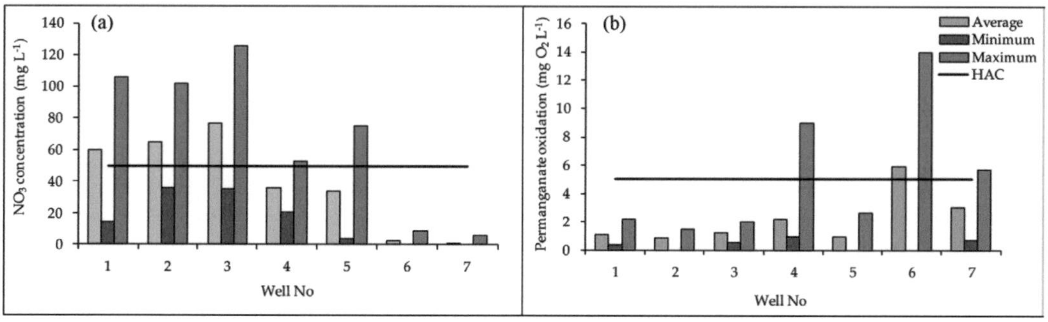

Figure 5. Nitrate concentration (a) and amounts of organic substance (b) in the drinking water wells. Wells 1–5 were identified in moraine sandy loam; wells 6 and 7 were identified in glacial lacustrine clay.

The study [42] shows that the type of soil (mineral and peat) and location of a well in respect of sinkholes have impacts on chemical composition of groundwater. Sum of ions (mineralization) in water of peat soil was a 1.1–1.2 times larger than in water of mineral soil (**Table 6**). According to the mineralization level and distribution of anions, the groundwater in a sinkhole (well 4) contained the largest amounts of hydro-carbonates (8.5 meq L^{-1}), calcium (5.9 meq L^{-1}), magnesium (2.9 meq L^{-1}) and potassium (0.26 meq L^{-1}); the distribution of cations was similar to the water samples collected from other wells ($Ca^{2+} > Mg^{2+} > Na^{+} > K^{+}$), but distribution of anions was different ($HCO_3^- > Cl^- > SO_4^{2-} > NO_3^-$ that is SO_4^{2-} ions were replaced by Cl^-).

Groundwater level fluctuation is influenced by precipitation and melting snow water amount. The most distinct fluctuations of groundwater level were observed in mineral soil where the difference was 3.2 m and in peat soil groundwater level fluctuations were 1.7 (well 2 installed near the stream) and 0.85 m (well 4 installed in a sinkhole).

The hot and dry summers, recurrent droughts, very uneven distribution of rainfall (over the year and in the territory) had a negative impact on groundwater resources [47]. Changes of groundwater level regime, especially seasonal, will not necessarily be useful for other elements of the environment to which groundwater is closely linked [48].

Changes of groundwater mineralization depend on its level fluctuation [42]. A larger ion sum results in lower groundwater level. A stronger correlation was determined between groundwater levels and chemical compounds (Ca^{2+}, Mg^{2+}, SO_4^{2-} and HCO_3^-) concentrations inducing groundwater mineralization. During the vegetation period (April–October), when water level was raised nearer by ground surface in mineral soil, nitrate concentration was increasing. The water levels subside for concentrations amount was insignificant. In peat, the groundwater level fluctuation and NO_3^- concentration have no connection.

Well No	K^+	Na^+	Ca^{2+}	Mg^{2+}	NH_4^+	HCO_3^-	SO_4^{2-}	Cl^-	NO_3^-	Ion sum
Precipitation										
	0.03	0.05	0.31	0.25	0.05	0.24	0.32	0.06	0.03	1.34
Groundwater in mineral soil										
1	0.05	0.31	5.2	2.4	0.02	6.7	0.62	0.26	0.33	15.9
3	0.04	0.21	5.0	2.1	0.04	6.7	0.21	0.18	0.12	14.6
Groundwater in peat										
2	0.11	0.44	5.2	2.4	0.45	8.0	0.16	0.09	0.01	16.8
4	0.16	0.23	5.9	2.9	0.24	8.5	0.20	0.28	0.02	18.4

Table 6. Average chemical composition (meq L^{-1}) of precipitation and groundwater between 1997 and 2002 [42].

6. Water protection

Based on the geological and hydrogeological studies, the karst area was divided into two parts. The first part is territory of active karst development. The second part is karst protection zone [4]. Depending on the degree of vulnerability of the karst terrain (thickness of overlying sediments, density of sinkholes, recharge-discharge characteristics and the amount of groundwater pollution), the active karst zone was subdivided into four land groups (**Figure 6**).

Each of these land groups has a different level of restriction imposed on agricultural activity within it [49]. These are land group 1 (from 5 to 20 sinkholes km^{-2})-perennial grass should compose at least 20% of crops. Fertilizers are limited up to 100 kg ha^{-1} total nitrogen of mineral fertilizer and manure. Land group 2 (from 21 to 50 sinkholes km^{-2})-perennial grass should compose at least 30% of crops. Fertilizers are limited up to 80 kg ha^{-1} total nitrogen of mineral fertilizer and manure. Land group 3 (from 51 to 80 sinkholes km^{-2})-perennial grass and grazing should compose at least 40% of crops. Fertilizers are limited up to 70 kg ha^{-1} total nitrogen of mineral fertilizer and manure. Land group 4 (>80 sinkholes km^{-2})—meadows and forest are allowed. Melliferous and drug plants should be grown. Fertilize can only use manure.

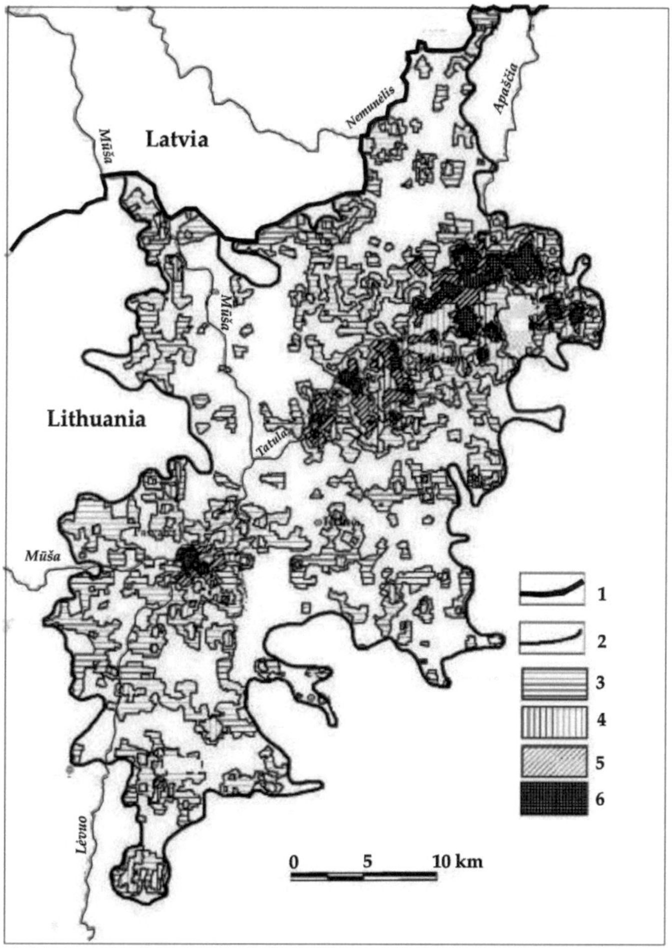

Figure 6. Scheme of distribution of karst area according to number of sinkholes: (1) state's border; (2) North Lithuanian karst region border; (3–6) land group (3—from 5 to 20 sinkholes km^{-2}; 4—from 21 to 50 sinkholes km^{-2}; 5—from 51 to 80 sinkholes km^{-2}; 6—>80 sinkholes km^{-2}) (according to the literature [49]).

In all land groups, a 5 or 10 m (according to the slope of the surface) width protection zone is required around each sinkhole.

In an alternative approach, the protection of karst water from pollution in the agriculture area is sustainability agriculture practise.

7. Summary

North Lithuanian karst region, as the mostly Lithuanian territory, is used for agricultural practise. The leaching of nutrients from agricultural fields is one of surface and groundwater pollution sources. Polluted drinking groundwater is dangerous for people's health. The salt-saturated water becomes more aggressive and intensifies karst process of the rocks lying below, which definitely causes a risk to people's property or event life.

Most chemical matter contained in the soil is water-soluble; therefore, hydrological regime changes in the ecosystem are closely related to changes in nutritive leaching. By means of drain, much nitrate nitrogen and other elements contaminating the stream (drainage receivers') water are leached out. The stream water mineralization was determined by soil texture, type of land use and fertilizers applied in the areas drained. Higher amount of nitrogen and phosphorus are leached out in winter and spring periods and at the onset of run-off after a dry period (due to the leaching of nitrogen that was not used during the dry period).

The bigger rivers have more intensive agitation processes in the flow and receive run-off from greater areas, while quality of small streams is under immediate impact of local factors and distinguished impact of particular factors in the catchments. The concentration of nitrate nitrogen is up to 2.1 times smaller in the rivers than it is in the stream.

Chemical composition of groundwater is mostly depended on the type of soil and location of a well in respect of sinkholes and meteorological conditions. In the dry period, peat-filled sinkhole water is at higher level and less polluted with nitrate nitrogen, chlorine and sulphate compounds.

Author details

Aurelija Rudzianskaitė

Address all correspondence to: Aurelija.Rudzianskaite@asu.lt

Institute of Water Resources Engineering, Faculty of Water and Land Management, Aleksandras Stulginskis University, Akademija, Lithuania

References

[1] Kaçaroğlu F. Review of groundwater pollution and protection in karst area. Water, Air, and Soil Pollution. 1999;**113**:337-356

[2] Satkunas J, Marcinkevicius V, Mikulenas V, Taminskas J. Rapid development of karst landscape in North Lithuania – monitoring of denudation rate, site investigation and implication for management. GFF. 2007;**129**:345-350

[3] Taminskas J. Generation and development of karst pits. Geografijos Metraštis. 1999;**32**: 194-203. (in Lithuanian)

[4] Paukštys B. Hydrogeology and Groundwater Protection Problems in Karst Region of Lithuania. Vilnius: Geology Society of Lithuania; 1996. p. 72

[5] Delle Rose M, Federico A, Parise M. Sinkhole genesis and evolution in Apulia, and their interrelations with the anthropogenic enviroment. Natural Hazards and Earth System Sciences. 2004;**4**:747-755

[6] Statistics Lithuania. Main country indicators [Internet]. 2014. Available from: http://db1.stat.gov.lt/statbank/ [Accessed: 14-11-2016]

[7] Narbutas V, Linčius A, Marcinkevičius V. Karst of the Devonian Rocks and the Problems of Environment Protection in North Lithuania. Vilnius: Agora; 2001. p. 191 (in Lithuanian)

[8] Povilaitis A, Lamsodis R, Bastiene N, Rudzianskaitė A, Misevičiene S, Miseckaite O, Gužys S, Baigys G, Grybauskiene V, Balevičius G. Agricultural drainage in Lithuania: A review of practices and environmental effects. Acta Agriculturae Scandinavica, Section B-Soil & Plant Science. 2015; **65**(Suppl.1):14-29

[9] Marcinkevicius V, Buceviciute S. Geological and hydrological conditions for the development of sulphate karst in North Lithuania. Geologija. 1986;**7**:104-121 (in Russian)

[10] Račkauskas V. The map of thickness and lithological composition of the cover of karstified rocks (North Lithuanian karst area). Lietuvos geologijos tarnybos 2003 metų veiklos rezultatai. 2004; 25-26. (in Lithuanian)

[11] Juodkazis V. Lithuanian karst: Hydrogeology and groundwater protection. Geologija. 1992;**13**;195 p. (in Lithuanian)

[12] Satkūnas J, Taminskas J, Dilys K. Geoindicators of changing landscapes – an example of karst development in North Lithuania. Geological Quarterly. 2006;**50**(4):457-464

[13] Bukantis A, Rimkute L, Kazakevičius S. The Variability of Climate Elements in the Lithuanian Territory. Vilnius: Institute of Geography; 1998. p. 171

[14] Statistics Lithuania. Database of Indicators by Region [Internet]. 2007. Available from: http://db1.stat.gov.lt/statbank/SelectVarVal/saveselections.asp [Accessed: 19-10-2016]

[15] Rudzianskaitė A, Šukys P. Drainage water mineralization of karst region soils and its impact on stream water quality. Vandens Ūkio Inžinerija. 2000;**11**(33):81-90

[16] Rudzianskaite A, Šukys P. Changes of nitrogen, phosphorus and potassium amounts in water and silts of the streams located in the karst zone. Water Management Engineering. 2005;**2**(5);86-92

[17] Rudzianskaite A. Leaching of nitrogen, phosphorus and potassium fro soil within the karst region. Vandens Ūkio Inžinerija. 2000;**11**(33):72-80 (in Lithuania)

[18] Morkunas V, Rudzianskaite A, Šukys P. Influence of agriculture on soil water quality in the karst region of Lithuania. Irrigation and Drainage. 2005;**54:**353-361

[19] Adomaitis T, Arbačiauskas J, Mažvila J, Morkūnas V, Rudzianskaite A, Šukys P. Investigations on chemical compounds leaching in the soils of active karst zone. Vandens Ūkio Inžinerija. 2005;**28(**48)1:94-101 (in Lithuanian)

[20] Rimkus E, Kažys J, Junevičiūtė J, Stonevičius E. The forecast of Lithuanian climate change in twenty-first century. Geografija. 2007;**43**:37-47

[21] Rudzianskaite A. Changes of drainage run-off hydrological regime in glacial lacustrine clays of karst zone. In: Selected Paper of the 8th International Conference Environmental Engineering; 19-20 May 2011. Vilnius: Technika Press; 2011. pp. 643-647

[22] Rudzianskaite A, Miseviciene S. Nitrate nitrogen leaching in different agroecosystems (in karst zone and Middle Lithuania). Journal of Water and Land Development. 2005;**9**:123-133

[23] Chomčenko R, Juodkazis V, Rudzianskaitė A, Taminskas J. Dynamics of nutrients in the geosystem of the karst region in North Lithuania. Vandens Ūkio Inžinerija. 2000;**11(**33):57-71 (in Lithuanian)

[24] Gailiušis B, Jablonskis J, Kovalenkovienė M. Lietuvos upės. Hidrografija ir nuotėkis. Kaunas: Lietuvos energetikos institutas; 2001. p. 407

[25] Vuorenmaa S, Rekolainen S, Lepisto A, Kenttamies K, Kauppila P. Losses of nitrogen and phosphorus from agricultural and forest areas in Finland during the 1980 and 1990. Environmental Monitoring and Assessment. 2002;**76:**213-248

[26] Kyllmar K, Carlsson C, Gustafson A, Ulen B, Johnsson H. Nutrient discharge from small agricultural catchments in Sweden. Characterisation and trends. Agriculture, Ecosystems & Environment. 2006;**115:**15-26

[27] Povilaitis A. Impact of agriculture decline on nitrogen and phosphorus loads in Lithuanian rivers. Ekologija. 2006;**1**:32-39

[28] Ruminaitė R, Šileika AS, Lukianas A. Analysis of pollution of the Musa catchment with total nitrogen. Ekologija. 2009;**55**(2):112-120

[29] Šileika AS. Direktyvos 91/676/EEC dėl vandenų apsaugos nitratų taršos, susidarančios dėl žemės ūkio veiklos, įgyvendinimo pasekmių įvertinimas. Kėdainiai: Mokslinė studija; 2001

[30] Purvinis O, Šidlauskas V, Šukys P. Nitrogen stream water: Stacionary and non-stacionary models. Water Management Engineering. 2006;**3**(6):16-22

[31] Stankevičiene R. Analysis of annual and seasonal total nitrogen pollution in the Mūša catchment. Selected Paper of the 9th International Conference Environmental Engineering; 23-23 May 2014. Vilnius; enviro.2014.094

[32] Taminskas J. Hidrologinis karstinio regiono ribų pagrindimo metodas. In: Šiaurės Lietuvos Karstinis Regionas. Geografiniai gamtonaudos aspektai. Vilnius : Geografijos institutas ; 2000. p.32-35 (in Lithuanian)

[33] Povilaitis A. Pagrindinių jonų koncentracijų karstinio regiono upėse panašumų ir skirtumų statistinis įvertinimas. Vandens Ūkio Inžinerija. 2003;**22**(44):5-14

[34] Gaigalis K, Šmitienė A. Žemdirbystės įtakos upių nuotėkio ir azoto koncentracijos kaitos analizė. Vandens Ūkio Inžinerija. 2004;**27**(47):45-50

[35] Ruminaite R, Rudzianskaite A. Change of water quality in the rivers and streams of Lithuania karst zone. Proceedings of the Fifth international Scientific Conference Rural Development' 2011; 24-25 November 2011; Akademija: Aleksandras Stulginskis University; 2011. pp. 332-336

[36] Vaitiekūnienė J, Virbickas T, Daunys D, Taminskas J. Paviršinių vandens telkinių išskyrimas ir apibūdinimas. In: Paukštys B, editor. Lietuvos Vandens Telkinių Būklė ir Ūkinės Veiklos Poveikis. Vilnius; 2011. pp. 17-120 (in Lithuanian)

[37] Taminskas J. Optimization of sinkholes protection. Geografijos Metraštis. 2002;**35:**87-94 (in Lithuanian)

[38] Rudzianskaite A. The change of water quality in the sinkholes in Lithuanian karst zone. Proceedings of 19th International Scientific Conference Research for Rural Development' 2013; 15-17 May 2013. Jelgava: Latvia University of Agriculture; 2013. pp. 153-160

[39] Minkevičius V. Stability map of the territory of North Lithuanian karst region. Lietuvos geologijos tarnybos 2010 metų veiklos rezultatai. 2011:60-61 (in Lithuanian)

[40] Šukys P, Šaulienė A. Investigation of groundwater regime and pollution in the vicinity of karst sinkholes. Vandens Ūkio Inžinerija. 2006;**29**(49):43-49 (in Lithuanian)

[41] Šukys P, Šauliene A. Researches of the groundwater flow regime through sinkholes and its quality. Vandens Ūkio Inžinerja. 2008;**33**(53):58-65. (in Lithuanian)

[42] Rudzianskaite A, Sukys P. Effects of groundwater level fluctuation on its chemical composition in karst soils of Lithuania. Environmental Geology. 2008;**56**(2):289-297

[43] Kazitėnas A. Šulinių vandenyje – nuodų kokteilis. 2007. Available from http://www.siaure.lt/article/articlearchiveview/244/2007/06/28 (in Lithuanian) [Accessed: 14-11-2008]

[44] Rudzianskaite A. Statistical evaluation of well-water quality. Vandens Ūkio Inžinerija. 2008;**34**(54):39-47 (in Lithuanian)

[45] Kutra S, Rudzianskaite A. Organic matter concentration in water of the karst zone dug wells. Proceedings of The Fourth International Scientific Conference Rural Development' 2009; 15-17 October 2009. Akademija; 2009. pp. 34-37

[46] Rutkovienė V, Kusta A, Kutra S, Česonienė L. Šachtinių šulinių vandens taršos sezoninė kaita. Vandens Ūkio Inžinerija. 2003;**22**:15-29

[47] Arustienė J, Kriukaitė J. Climate cange impact on groundwater resources. In: Kadūnas K, editor. Lietuvos Požeminio Vandens Monotoringas 2005-2010 m. ir kiti hidrogeologiniai darbai. Vilnius: Lietuvos Geologijos Tarnyba; 2011. pp. 84-89 (In Lithuanian)

[48] Giedraitienė J. Change of groundwater level regime associated with meteorological condition). In: Kadūnas K, editor. Lietuvos požeminio vandens monotoringas 2005-2010 m. ir kiti hidrogeologiniai darbai. Lithuania: Lietuvos Geologijos Tarnyba, Vilnius; 2011. pp. 13-22 (In Lithuanian)

[49] Dėl Šiaurės Lietuvos karstinio rajono ribų ir intensyvaus karsto žemių grupių nustatymo [Internet]. 2008. Available from: https://www.e-tar.lt/portal/lt/legalAct/TAR.59906C3193F6 [Accessed: 8-12-2016]

Chapter 6

ERT and the Location of Mining Cavities in Anisotropic Media: A Field Example

Manuel Matias and Fernando Almeida

Additional information is available at the end of the chapter

http://dx.doi.org/10.5772/intechopen.68475

Abstract

Often cave location requires the use of surface indirect techniques, such as geophysical methods. In particular, electrical methods have been applied to cavity exploration with evident success. However, as any other indirect methods, the use of these techniques has advantages and disadvantages. Cavities may be too small, too deep or masked by local geology to be detected. Nevertheless, indirect methods provide non-invasive, low cost and fast techniques to carry out the reconnaissance of an area where the presence of cavities is suspected. Complex geological conditions and formations anisotropy can induce strong orientational variation on ground resistivity measurements and, therefore, mask the presence of caves. Herein a field study in an old mining area demonstrates that 2D resistivity data—electrical resistivity tomography (ERT)—can be strongly affected by local anisotropy that masks the presence of cavities in ERT data modelling. In these cases, specific field strategies must be considered to overcome misleading interpretations and modelling, so that, meaningful results are obtained, uncertainty and interpretation ambiguity are reduced and the correct diagnosis of caves is accomplished.

Keywords: exploration, cavity, orientation, ERT, anisotropy

1. Introduction

The importance of cavities in engineering, mining, environment, archaeology, tourism, etc. has long been acknowledged. Caves have natural origin (lava, karst or other dissolution processes) or can originate from anthropic activities, such as mining, engineering excavations and archaeological features.

Knowledge of cave location and extension is very important during the construction and maintenance of infrastructures (tunnels, highways, railways, sewage systems) and when urban areas extend to old mining areas or karst regions. In fact, risks of collapse and pollutants transportation can lead to unwanted and hazardous events. Caves are also used as reservoirs, deposits of gas, hazardous and toxic residues and, therefore, cavity monitoring and four-dimensional (4D) studies must be considered for safety reasons.

Cavity location is often done by surface mapping, documents, local and oral descriptions. Therefore, before engaging in expensive and comprehensive exploration programmes, it is important to review all the relevant available information. However, often local records are difficult to obtain, particularly in old mining areas. It is also possible that no information is available in regions with no evidence of caves at the surface or in cases of fast cavity development, such as sinkholes.

Often invasive exploration methods (excavation, drilling) are used. These invasive methods must be carefully planned to reach the targets at a suitable cost and, furthermore, operations must not interfere and damage the caves [1].

Other approach uses indirect investigation techniques from the surface that is geophysical exploration methods. However, the efficiency of these methods depends, among other factors, on the contrast between the physical properties of the cavities and those of the surrounding media [2].

Fortunately, in most cases there is a contrast between the velocity of seismic wave propagation, density, electrical and magnetic properties of the cavities and those of the rock formations where they are installed [3] and, thus, geophysical exploration methods can be used and adapted to cavity detection and location [4]. However, the relation between the dimensions of the cavities and their depth can also be a major limitation factor for the use of geophysics. In fact, cavities can be too small or too deep to be detected in spite of a large contrast in physical properties [5].

Since these limitations are overcome, the main objective of the use of geophysical methods in cavity location is to provide information and restrict the area to investigate by direct methods, to guide later exploration operations and to diminish costs while preserving the targets [6].

In this contribution, the application of geophysical methods in cavity exploration is focused on the use of 2D resistivity methods—electrical resistivity tomography (ERT).

There many examples of the use of 2D and 3D ERT in cavity location [7, 8]. As in any other geophysical method, the success of cavity detection by resistivity methods depends on factors such as depth, size and contrast between the resistivity of the cavity and that of the surrounding media [5]. Cavities can be more conductive or more resistive then the rocks that surround them. Cavities filled with water are more conductive then the surrounding media. However, when empty they are more resistive as air is not an electricity conductor. Thus, in the first case, cavity response to resistivity methods is a conductive anomaly whilst, in the second case, the response will be a resistive anomaly.

The nature of the surrounding media is another factor to consider. Usually, resistivity fieldwork consists on recording data in one unique direction. Therefore, two-dimensional effects arising from local geology such as contacts, schistosity, etc., can induce orientational variation on surface resistivity data and complicate cavity detection.

http://dx.doi.org/10.5772/intechopen.68475

In the past, resistivity fieldwork was a slow operation. Nowadays, the development of automated equipment allows the fast acquisition of large field data sets. This enables the production of high-resolution images of the ground and answers the engineers' and planners' needs for fast, non-intrusive and high-resolution methods to detect underground targets such as cavities.

This work will give a brief introduction to the resistivity methods and to the field techniques necessary to carry out an ERT. Then an approach to the resistivity behaviour in anisotropic media is presented. Finally, a case study concerning the location of mining cavities in anisotropic regions is discussed. The ambiguity and uncertainty in the ERT interpretation will be addressed and field strategies to reduce or overcome those limitations are also presented.

2. Basic theory of resistivity methods

Resistivity measurements are traditional geophysical methods. There are extensive textbooks presenting the theory of electrical methods [9] and their use in engineering and environmental investigations [10].

In broad terms, resistivity methods consist on passing a DC current into the ground using two current electrodes and measuring the generated electrical potential between two potential electrodes, **Figure 1**.

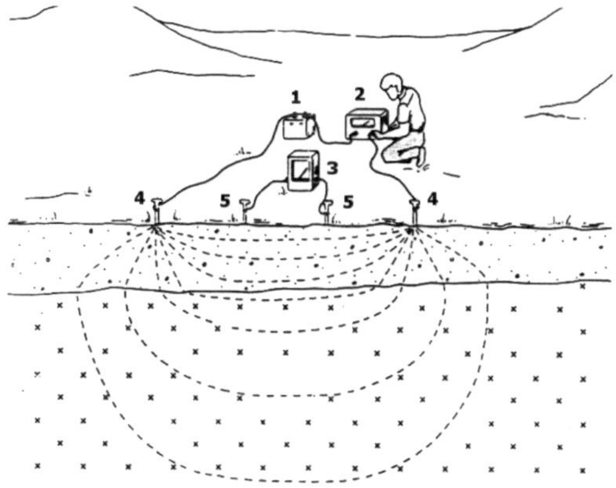

Figure 1. Resistivity fieldwork; 1, battery; 2, ammeter; 3, voltmeter; 4, current electrodes; 5, potential electrodes; dashed lines, current lines in the ground.

In the past, resistivity field operations were slow and tedious but the development of computer controlled multi-electrode resistivity metres and cables, **Figure 2**, enables fast field operations and the recording of large data sets.

Usually, field measurements use four in-line electrode arrays. The commonest in-line electrode arrays are the Wenner (left of **Figure 3**) and Schlumberger (right of **Figure 3**).

Figure 2. Resistivity equipment: left, automated resistivity metre and cables; right, stainless steel electrodes.

Figure 3. Resistivity arrays: left, Wenner; right, Schlumberger.

In the Wenner array, the electrodes are equally spaced, 'a', whilst in the Schlumberger array the distance AB, between the current electrodes, is larger than the distance between the potential electrodes MN. Once the potential difference, V, and the current, I, are measured the resistivity of the ground, ρ, is

$$\rho = 2\pi a \frac{\Delta V}{\mathrm{I}} \quad \text{for the Wenner array} \tag{1}$$

and

$$\rho = \frac{\Delta V}{\mathrm{I}} \pi \frac{(b+a)}{a} \quad \text{for the Schlumberger array} \tag{2}$$

The construction of an ERT requires the use of Wenner and Schlumberger arrays of different sizes along an acquisition line of electrodes, **Figure 4**.

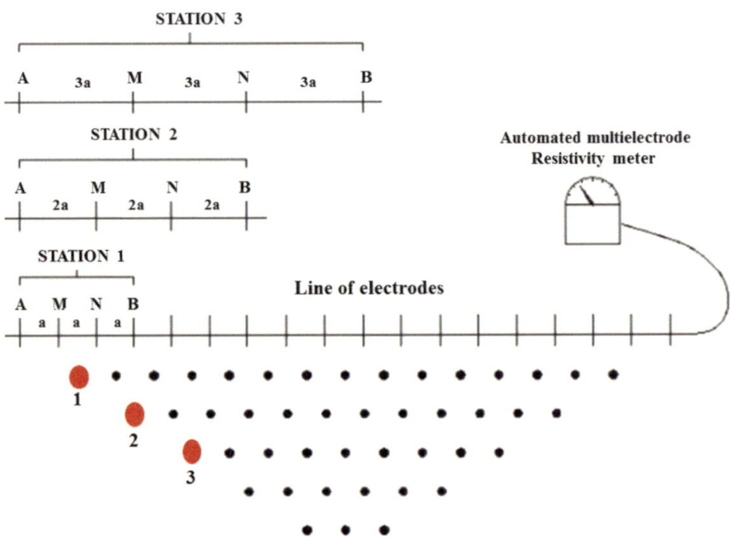

Figure 4. Construction of an electrical resistivity tomography—ERT.

As depicted in **Figure 4**, the first line is obtained by moving the electrode array, using an electrode spacing 'a', along the line of electrodes. The measurements are plotted in accordance with the centre of the array and the spacing 'a' used, first line in **Figure 4**. The first measurement corresponds to the red dot 1, station 1.

When line 1 is finished, field measurements resume with line 2. Now, the electrode array spacing is '2a' and data are plotted against the position of the centre of the array and the new spacing '2a'. The first measurement corresponds to the red dot 2, station 2.

The procedure continues until reaching the number of lines required to obtain an image—ERT—of the ground.

To optimize field measurements, it is usual to combine Wenner and Schlumberger arrays. The complete procedure is controlled by an automated multi-electrode resistivity metre previously programmed to carry out the complete sequence of field measurements and to store all field information.

Once all data are stored, they can be modelled using appropriate software [11, 12] and detailed information about the ground is obtained.

Often cavities are installed in heterogeneous and complex media. In these cases, resistivity measurements can vary with the orientation of the line of electrodes, as shown in **Figure 4**. In such cases, the orientational variation of resistivity data can hinder the location of cavities and, possibly, mask their presence.

In the presence of anisotropic media, steeply dipping schists are a good approximation, it is usual to consider two resistivities: one ρ_l the longitudinal resistivity, in the direction of the strike, and another ρ_t, transverse resistivity, perpendicular to the strike [13], **Figure 5**.

As the longitudinal and transverse resistivity values are different, it is possible to define a mean resistivity ρ_m:

$$\rho_m = \sqrt{\rho_t \rho_l} \quad (3)$$

and an anisotropy coefficient, λ, the square root of the ratio between ρ_t and ρ_l:

$$\lambda = \sqrt{\rho_t / \rho_l} \quad (4)$$

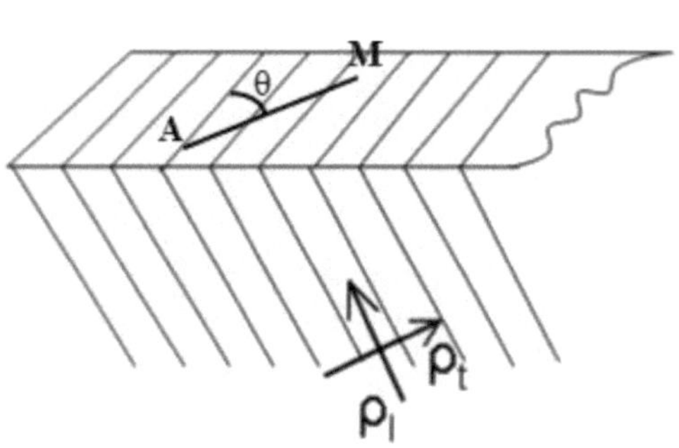

Figure 5. Anisotropic medium: ρ_l, longitudinal resistivity; ρ_t, transverse resistivity; A, current electrode; M, potential electrode; θ, array orientation.

In the field, resistivity values can vary largely with the orientation θ of the line of electrodes, **Figure 5**, and the coefficient of anisotropy can reach values larger than 2 [9, 13].

The influence of anisotropy on resistivity measurements has been investigated [14, 15] and, herein, a case study on the location of old mining cavities in anisotropic media will be presented [16].

3. Cavity detection in anisotropic regions: a field example

Slates mining in Valongo (41°11′N, 8°30′W, 30 km North of Oporto, North Portugal) started in the nineteenth century. Since then many underground works were abandoned and no records are left about their whereabouts and extension. Slates of economic value occur in steeply dipping Ordovician schists where strong orientation effects of physical properties and anisotropy occur.

Increasing urban development pressure requires information about the position and extension of mining chambers but, documents about older works are scarce and field interventions demand the use of fast and non-invasive methods that is geophysics.

Electrical tomography techniques (ERTs) using the Wenner-Schlumberger array were used to investigate the dimensions of an old mining chamber. Today, the only evidence of the chamber at the surface is the shaft and the possible lateral limits of the underground works, depicted by the dashed line in **Figure 6**.

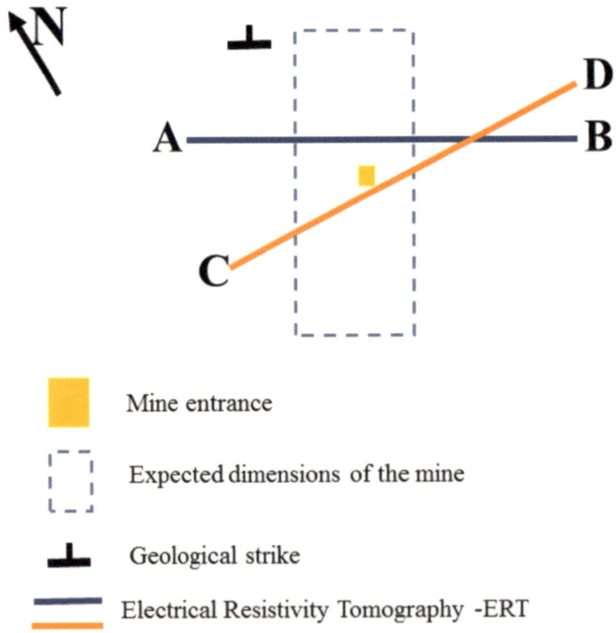

Figure 6. Mine chamber and ERT location in Valongo.

Figure 6 also shows the mine entrance (central gray rectangle), the position of two ERTs, AB and CD, and the geological strike. Inspection of the shaft revealed that the mine is flooded and the groundwater level is very close to the surface. Thus, a large contrast between the cavity (conductive) and the surrounding rock (resistive) is expected. So a conductive anomaly is expected.

Because of access difficulties, a first ERT, AB in **Figure 6**, was carried out parallel to the geological strike and, in these conditions, values corresponding to ρ_l should have been measured. Data were inverted using the software RES2Dinv [12], and the model is shown in the left of **Figure 7**.

The model shows a resistive layer overlaying a conductive formation. The boundary, shown by the dashed line in the left of **Figure 7**, is at the same depth of the groundwater in the chamber entrance.

As the ERT orientation is parallel to the strike, the modelled resistivity should correspond to the longitudinal resistivity, ρ_l. Therefore, longitudinal resistivity values must be higher above the groundwater level. However, they must decrease sharply below that level as a response to the water in the chamber and in the schists' foliation. Hence, this boundary is interpreted as the groundwater level.

Below this boundary, no more relevant information is obtained from the ERT model as only a conductive medium is shown.

Thus, tomography AB does not give evidence of the chamber and, if only this ERT was carried out, the location of the chamber would have been completely missed.

Therefore, it was decided to conduct a second ERT, CD in **Figure 6**. Because of access difficulties, CD orientation was at approximate 45° to the geological strike. The same field technique was used and model results are depicted in the right of **Figure 7**. In this case, a conductive body, bounded by the dashed lines, is clearly shown and stands out from the resistive surrounding rocks.

At shallow depths, a resistive layer is depicted with the same thickness in both ERTs. Bearing in mind the previous discussion, this layer corresponds to the resistive ground above the groundwater level.

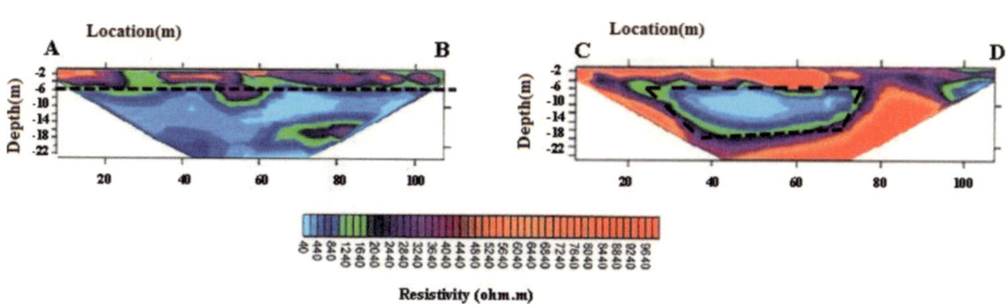

Figure 7. ERT models in Valongo: left, AB (dashed line, groundwater level); right, CD (dashed line, interpreted cavity dimensions).

As the chamber is filled with water, the conductive anomaly should correspond to the mining chamber and, in fact, the proposed lateral boundaries match the expected limits of the dashed line in **Figure 6**. The higher resistivity values recorded away from the conductive anomaly refer to the schists' resistivity as measured with this ERT orientation.

Because of the orientation used, modelled resistivity values in CD do not correspond to the transverse resistivity of the schists. Therefore, it is not possible to calculate the anisotropy coefficient, but an indication of field anisotropy values can still be computed by calculating the square root of the ratio between the modelled resistivities for CD and AB.

Thus, a further section was constructed depicting the square root of the ratio between the resistivities modelled in the two ERTs.

Figure 8 shows ratio values varying from 1 to more than 3.4. Thus, although this is not the anisotropy coefficient, field conditions are highly anisotropic.

The high resistive shallow layer, above the groundwater level, should correspond to ratio values close to 1. In this case, resistivity values modelled for AB and CD refer to a 'dry' or non-saturated area and, thus, should be similar. Hence, the dashed red line in **Figure 8** must correspond to the groundwater level as shown in **Figure 7** and, above this boundary, ratio values vary from 1 to 1.4. These values are considered to correspond to a low anisotropic medium [13, 17].

The central area of the section displays resistivity ratio values near 1 (or less than 1). This area corresponds to the position of the chamber depicted in the right of **Figure 7**. As the chamber is filled with water, resistivity values modelled for AB and CD ERTs should be similar as they correspond to data in the conductive chamber region.

It must be noted that ratio values less than 1 are depicted in **Figure 8**. This behaviour has been registered where structural anisotropy prevails [13, 17], such as in the vicinity of interfaces separating media with high resistivity contrast. In these circumstances, there is a rotation of the anisotropy axis and the so-called oblate anisotropy is observed [13]. This should be the

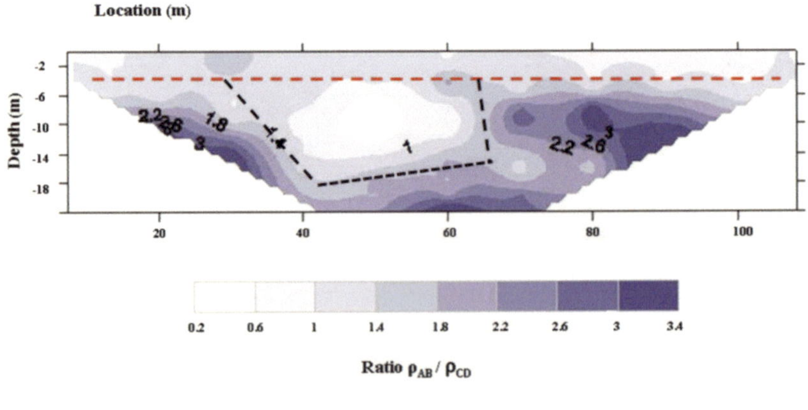

Figure 8. Ratio between modelled resistivities in Valongo; dashed line, proposed cavity limits.

case in the field example as the contact between the conductive chamber and the highly resistive surrounding rock corresponds to an interface separating two media with a very high resistivity contrast.

Away from the chamber position, ratio values increase as it should be expected. Therefore, one direction—AB—corresponds to the resistivity in the foliation planes (lower values), but the other direction—CD—is closer to the transverse resistivity (higher values).

There are many examples of the use of conventional ERT techniques in the location of mining cavities with evident success. However, the influence of formation anisotropy is seldom addressed and can mask ERT response. In this case, different field ERT orientations must be used to avoid misleading interpretations.

Often field space is restricted and only one ERT can be carried out. In this case, ERT orientation must be rather different from the strike of the geological formations and, in optimal conditions, ERT orientation should be perpendicular to the strike.

It is also possible to use electrode arrays that take into account orientation effects and anisotropy. Therefore, sets of linear arrays covering a wide range of directions can be used [14]. Alternatively, arrays of different geometry such as the square array have also been proposed [13].

However, the use of alternative arrays demands space and easy field access conditions and this requirement can be difficult to meet in urban areas.

4. Conclusions

Geophysical methods are a powerful tool for cavity location. They provide fast, economic, automated, non-invasive techniques that offer relevant information to restrict areas of interest and, thus, to guide more expensive direct exploration methods. As non-invasive methods, they do not require excavation or drilling and thus can be adapted to operate in urban areas.

These methods are indirect techniques as they measure the difference in physical properties between the cavities and surrounding media and not the properties of the cavities themselves. Fortunately, most of the times, there is a contrast between the physical properties of the cavities and those of the surrounding rocks. However, the size and depth of the cavities can be a limiting factor for the use of geophysical methods.

The complexity of the geology formations where the cavities are installed can be another limitation factor as demonstrated in this case study. Therefore, orientational effects and formation anisotropy can mask cavity response inducing ambiguity and uncertainty in the interpretation. In this case, field survey design must be carefully planned to overcome or reduce orientational effects in the final interpretation, to avoid misleading interpretations and the use of alternative non-linear arrays must be considered.

Acknowledgements

The authors wish to thank the Portuguese Science Foundation, project UID/GEO/04035/2013.

Author details

Manuel Matias* and Fernando Almeida

*Address all correspondence to: mmatias@ua.pt

GeoBioTec, Department of Geosciences, University of Aveiro, Aveiro, Portugal

References

[1] Nielsen D. Practical Handbook of Environmental Site Characterization and Ground Water Monitoring. CRC Press, Taylor and Francis Group, Boca Raton, Fl,US; 2006. 1281 p. ISBN-13:978-1-4200-3224-6

[2] Bates E.R. Detection of subsurface cavities. US Army Engineer Waterways Experiment Station,Vicksburg, Miss, US; 1973. Misc. Paper, S-73-40

[3] Elawadi E. Cavity Detection using Geophysical Methods: Evaluation and Applications. VDM Verlag Dr. Muller; 2010. 152 p. ISBN-13:978-3639297454.

[4] Butler D. Detection and characterization of subsurface cavities, tunnels and abandoned mines. Near-Surface Geophysics and Human Activity. 2008;578-584.

[5] James P, Ferreira P. Geophysical modeling of typical cavity shapes to calculate detection probability and inform survey design. Journal of Environmental and Engineering Geophysics. 2013;297-316

[6] Benson R, Glaccum R, Noel M. Geophysical Techniques for Sensing Buried Wastes and Waste Migration. Las Vegas: Environmental Monitoring Systems Lab, US Environmental Protection Agency; 1982. 236 p.

[7] Van Schoor M. Detection of sinkholes using 2D electrical resistivity imaging. Journal of Applied Geophysics. 2002;393-399. DOI: 10.1016/S0926-9851(02)00166-0

[8] Martinez-Lopez J, Rey J, Dueñas J, Hidalgo C, Benavente J. Electrical tomography applied to the detection of subsurface cavities. Journal of Cave and Karst Studies. 2013;28-37. DOI: 10.4311/2011ES0242

[9] Keller G, Friscknecht F. Electrical methods in geophysical prospecting. In: International Series in Electromagnetic Waves. Vol. 10. Pergamon Press; 1979. 523 p. ISBN:0-08-011525-X

[10] Reynolds J. An Introduction to Applied and Environmental Geophysics. 2nd ed. Wiley and Sons; 2011. 712 p. ISBN-13: 978-0471485360

[11] DC2DInvRes – Direct Current 2D Inversion and Resolution [Internet]. 2007. Available from: http://dc2dinvres.resistivity.net/ [Accessed: 4 January 2017]

[12] Loke M, Barker R. Rapid least-squares inversion of apparent resistivity pseudo sections using a quasi-Newton method. Geophysical Prospecting. 1996;131-152.

[13] Habberjam G. Apparent resistivity observations and the use of square array techniques. Geoexploration Monographs No. 9. Berlin: Geopublication Associates; 1979. 152 p. ISBN-3443130135 9783443130138

[14] Matias M. Electrical strike imaging and anisotropy diagnosis from surface resistivity measurements. Near Surface Geophysics. 2008;49-58. DOI: 10.3997/1873-0604.2007031

[15] Bolshakov DK, Modin IN, Pervago EV, Shevnin VA. Anisotropy effects investigations by resistivity methods in some inhomogeneous media. In: 57th EAGE Conference and Exhibition Extended Abstracts, P034. Glasgow; 1995. 5p.

[16] Matias M. ERT interpretation – Now you see it, now you don't, now what? In: 18th European Meeting of Environmental and Engineering Geophysics, Paris, France, September 2012. 5 p. DOI: 10.3997/2214-4609.20143344

[17] Matias M, Habberjam G. The effect of structure and anisotropy on resistivity measurements. Geophysics. 1986;964-971.